Lecture Notes of the Institute for Computer Sciences, Social Informatics and Telecommunications Engineering 248

More information about this series at http://www.springer.com/series/8197

Jason J. Jung · Pankoo Kim
Kwang Nam Choi (Eds.)

Big Data Technologies and Applications

8th International Conference, BDTA 2017
Gwangju, South Korea, November 23–24, 2017
Proceedings

 Springer

Editors
Jason J. Jung
Chung-Ang University
Seoul, Korea (Republic of)

Pankoo Kim
Chosun Unversity
Gwangju, Korea (Republic of)

Kwang Nam Choi
Department of Computer Engineering
Chung-Ang University
Seoul, Korea (Republic of)

ISSN 1867-8211   ISSN 1867-822X   (electronic)
Lecture Notes of the Institute for Computer Sciences, Social Informatics
and Telecommunications Engineering
ISBN 978-3-319-98751-4   ISBN 978-3-319-98752-1   (eBook)
https://doi.org/10.1007/978-3-319-98752-1

Library of Congress Control Number: 2018958767

This Springer imprint is published by the registered company Springer Nature Switzerland AG
The registered company address is: Gewerbestrasse 11, 6330 Cham, Switzerland

Preface

The emergent area of big data brings together two aspects of data science and engineering: tools and applications. Theoretical foundations and practical applications of big data set the premises for the new generation of data analytics and engineering.

BDTA 2017 was the eight event in the series and was hosted by Chosun University in Gwangju, South Korea, during November 23–24, 2017. The material published in this book consists of 16 contributions of BDTA participants.

We would like to thank to Prof. Imrich Chlamtac, editor of *Lecture Notes of the Institute for Computer Sciences, Social Informatics and Telecommunications Engineering* (LNICST), and members of the Steering Committee for their kind support and encouragement in starting and continuing the BDTA Series. We thank the invited speaker, Dr. Chun-Wei Tsai, for the interesting lecture. Finally, we appreciate the efforts of the local organizers on behalf of Chosun University for organizing and hosting BDTA 2017.

February 2018

Jason J. Jung
Pankoo Kim
Kwang Nam Choi

Organization

Steering Committee Chair

Imrich Chlamtac   EAI/Create-Net, Italy

Steering Committee Members

Costin Badica	University of Craiova, Romania
David Camacho	Universidad Autonoma de Madrid, Spain
Jason J. Jung	Chung-Ang University, South Korea
Attila Kiss	Eötvös Loránd University, Hungary
Ngoc Thanh Nguyen	Wroclaw University of Technology, Poland
Le Anh Vu	Nguyen Tat Thanh University, Vietnam

Organizing Committee

General Co-chairs

Kwang Nam Choi	Chung-Ang University, South Korea
Pankoo Kim	Chosun University, South Korea

Technical Program Committee Co-chairs

Jason J. Jung	Chung-Ang University, South Korea
Chang Choi	Chosun University, South Korea

Web Chair

Hoang Long Nguyen   Chung-Ang University, South Korea

Publicity and Social Media Chair

Francesco Piccialli   University of Naples, Italy

Publications Chair

Jason J. Jung   Chung-Ang University, South Korea

Workshops Chair

Duc T. Nguyen   Vietnam Maritime University, Vietnam

Local Chair

Chang Choi   Chosun University, South Korea

Conference Manager

Alzbeta Mackova   European Alliance for Innovation

Technical Program Committee

Aniello Castiglione	University of Salerno, Italy
Antonio Gonzalez-Pardo	Universidad Autonoma de Madrid, Spain
Attila Kiss	Eötvös Lorán University, Hungary
Bahtijar Vogel	Malmö University, Sweden
Cesar Analide	University of Minho, Portugal
Chang Choi	Chosun University, South Korea
Chao-Tung Yang	Tunghai University, Taiwan
Christopher Lueg	University of Tasmania, Australia
Costin Badica	University of Craiova, Romania
David Bednarek	Charles University in Prague, Czech Republic
David Camacho	Universidad Autonoma de Madrid, Spain
Davide Carneiro	University of Minho, Portugal
Filip Zavoral	Charles University in Prague, Czech Republic
Fábio Silva	University of Minho, Portugal
Giancarlo Fortino	University of Calabria, Italy
Giuseppe Mangioni	University of Catania, Italy
Hector Menendez	University College London, UK
Igor Kotenko	St. Petersburg Institute for Informatics and Automation of the Russian Academy of Sciences (SPIIRAS), Russia
Juwook Jang	Sogang University, South Korea
Jason J. Jung	Chung-Ang University, South Korea
José Machado	University of Minho, Portugal
José Neves	University of Minho, Portugal
Juan Pavón	Complutense University of Madrid, Spain
Lars Braubach	Bremen University of Applied Sciences, Germany
Leon S. L. Wang	National University of Kaohsiung, Taiwan
Massimo Ficco	University of Naples, Italy
Michal Wozniak	Wroclaw University of Technology, Poland
Mirjana Ivanovic	University of Novi Sad, Serbia
Mohammad Hajiesmaili	The Chinese University of Hong Kong, SAR China
O-Joun Lee	Chung-Ang University, South Korea
Pankoo Kim	Chosun University, South Korea
Paulo Moura Oliveira	UTAD University, Portugal
Paulo Novais	ALGORITMI Centre/DI - University of Minho, Portugal

Phan Cong Vinh	Nguyen Tat Thanh University, Vietnam
Radu-Emil Precup	Politehnica University of Timisoara, Romania
Seung Park	Dankook University, South Korea
Tzung-Pei Hong	National University of Kaohsiung, Taiwan

Contents

S/W Engineering and E-Commerce

Social Media and Health Care

Privacy and Security

Study on the Transaction Linkage Technique Combining the Designated Terminal

Kyungroul Lee[1], Habin Yim[2], Insu Oh[3], and Kangbin Yim[3(✉)]

[1] R&BD Center for Security and Safety Industries (SSI),
Soonchunyang University, Asan, South Korea
carpedm@sch.ac.kr
[2] Center for Information Security Technologies (CIST), Korea University,
Seoul, South Korea
habin103@korea.ac.kr
[3] Department of Information Security Engineering, Soonchunhyang University,
Asan, South Korea
{catalyst32,yim}@sch.ac.kr

Abstract. While the scale of markets for Internet banking and e-commerce is growing, the number of financial markets using the Internet is increasing. However, there are a large number of hacking incidents against Internet banking services. For this reason, a countermeasure to improve the security of online identification is required. Security and authentication mechanisms applied to financial services such as Internet banking currently do not ensure security. In this paper, a transaction linkage technique combining a designated terminal is proposed to solve this fundamental problem, and the technique improves security for online identification mechanisms because it is possible to counteract all existing security threats. We consider that the security of Internet banking services will be enhanced by utilizing the proposed technique.

Keywords: Transaction linkage technique · Designated terminal
Internet banking service

1 Introduction

While the scale of markets for Internet banking and e-commerce is growing, the exchange of goods and services via the Internet has been established as a large part of the international economy. Although a variety of secure techniques is applied in the process of building these systems, hacking incidents on Internet banking services have occurred [1]. The damage resulting from such hacking and eavesdropping incidents on telebanking systems is continuous. In addition to general security applications, security techniques for online financial services are needed to ensure security requirements such as confidentiality, integrity, availability and non-repudiation [3]. Various cryptography-based mechanisms have been developed to satisfy these requirements over the past few decades [2], and their effectiveness was sufficiently verified by utilizing proven mathematical tools. Nevertheless, most security problems emerge during the process or in the environment of the security application rather than in the cryptography-based

© ICST Institute for Computer Sciences, Social Informatics and Telecommunications Engineering 2018
J. J. Jung et al. (Eds.): BDTA 2017, LNICST 248, pp. 3–8, 2018.
https://doi.org/10.1007/978-3-319-98752-1_1

technology. Hence, we understand the need for studies that focus on research to find vulnerabilities other than the cryptography-based technology and to counteract these vulnerabilities properly.

Online identification methods do not ensure security owing to existing and new security threats against the identification methods previously mentioned. Hence, we propose a transaction linkage technique combining a designated terminal to solve the problem of exposure to threats on Internet banking services. In the case where existing transaction linkage techniques have been used, these techniques can solve exposure problems from security threats, which are analyzed in this paper. Nevertheless, the techniques ca be abused when the transaction linkage device is stolen; that is the biggest problem of possession-based identification methods, and the linkage code is exposed because the code is inputted by keyboard. In addition, the techniques do not satisfy mutual-authentication because they are authenticated one-way, and do not satisfy non-repudiation of financial institutions for a user because the transaction history is stored in the financial institutions. Therefore, we propose a new transaction linkage technique combining a designated terminal, that is an approved transaction-only designated terminal, to solve the above problems. The proposed technique deals with transaction-only designated terminal registered by the user. This technique counteracts when the device is stolen, supports non-repudiation by storing transaction history into the transaction linkage device, and provides mutual-authentication. Hence, the technique can counteract most existing security threats by applying the above functions; therefore, we consider that it improves the security of online identification methods for Internet banking services.

2   Related Works

The transaction linkage technique is shown in Fig. 1. When a user inputs transaction information such as account number, transfer amount, and so on into the transaction linkage device, the device displays the linkage code (verification code) generated based on the sharing key between the Internet banking server and the device. Next, the user inputs the displayed code into the web browser; then, the code is transferred to the Internet banking server.

The existing transaction linkage technique, however, can be abused when the device is stolen and the linkage code can be exposed because the code is inputted from the keyboard. Moreover, the server only authenticates the device as one-way authentication, not mutual authentication that authenticates between the server and the device; the technique does not support non-repudiation because the transaction history is only stored in the financial institutions.

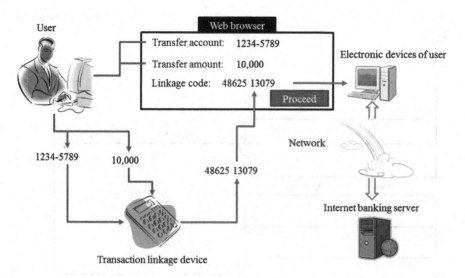

Fig. 1. Operational process of transaction linkage technique

3   Proposed Transaction Linkage Technique Combining a Designated Terminal

In this paper, we propose a transaction-linkage technique combined with a designated terminal to improve the security against the problems described above. The proposed technique is used only with a designated terminal; hence, the technique can counteract when the linkage device is stolen, and mutual authentication is provided between the server and the device. Moreover, a generated linkage code is transferred directly to the server, not inputted from the keyboard, and this technique provides non-repudiation by storing transaction history within the device. The operational process of the proposed technique is shown Fig. 2.

Stage 1.   In the registration process, a user applies service of designated terminal device (SDTD) to the financial institutions and registers the hardware unique information (HWUI) of electronic devices that the user wants to register as transaction linkage devices.

Stage 2.   After applying SDTD, the user identities himself or herself by offline authentication to visit the financial institutions directly, and he or she obtains the transaction linkage device after the offline authentication. The server and the transaction linkage device share a seed value for generating an encryption/decryption key, and time synchronization is applied in this stage.

Stage 3.   The user begins the financial transaction by accessing financial transaction sites in the authentication process.

Stage 4.   The user and financial institutions share a session key to establish a secure channel in the network communication.

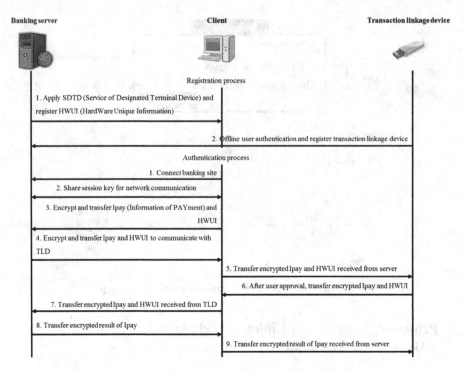

Fig. 2. The operational process of proposed technique

Stage 5.   The user sends transfer information, which is encrypted inputted transaction information and hardware unique information of the designated device, to the server.

Stage 6.   The server sends encrypted transaction information received and hardware unique information based on the shared encryption key between the server and the transaction linkage device to communicate with the device.

Stage 7.   The user authenticates the server based on the received information, and then sends the encrypted transaction information and hardware unique information to the transaction linkage device.

Stage 8.   The transaction linkage device displays an extra module such as a LCD (Liquid Crystal Display) panel for user recognition by decrypting the transaction information received, and the user approves the transaction after confirmation that the transaction information is correct. When the transaction is approved, the device sends encrypted transaction information and hardware unique information approved by the user to the server. If the transaction information is not correct, the transaction information mingled with random information is sent to the server in order to disrupt the communication process.

Stage 9.   The user sends information received from the device directly to the server.

Stage 10.   The server decrypts the transaction information received from the device and detects manipulation by comparing decrypted transaction information with received transaction information from the user. If the compared result is correct, this transaction is approved properly and the server sends the encrypted result, which is the processed transaction result.

Stage 11.   The user sends the received transaction result to the transaction linkage device.

Stage 12.   The transaction linkage device displays the received decrypted transaction result, and when the user finally confirms a transaction result, the transaction result is stored inside the device for non-repudiation

The server and transaction linkage device generate the encryption/decryption key based on a generated time stamp based on shared seed value and time synchronization. The generated key comprises a hash-chain type application based on time stamping to prevent encryption/decryption of transaction information and hardware unique information based on the same encryption/decryption key. Moreover, the session key for network communication in Stage 4 is changed in every session to prevent a replay attack, and a fixed password method, a one of the knowledge-based identity verification method, is applied to the transaction linkage device to improve transaction security. In addition, the transaction linkage device is flexible and can be applied to a variety of devices by communicating wirelessly or by a connector that can be inserted into the PC or mobile device. In terms of the safety of the proposed technique, the technique is safe from the debugging and reverse engineering attack because the transaction information is encrypted and decrypted in the server and the device inside by generating a key using a hash-chain type application based on shared seed value and time stamp. Therefore, the information is safe during sending and receiving processes between the network,

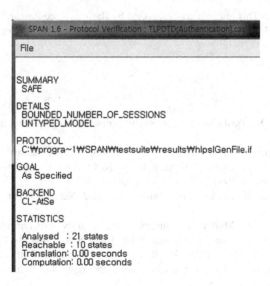

Fig. 3. Assessment result of security for proposed protocol by AVISPA

server, host, and the device. Moreover, a communication process between the host and the device is concealed, not exposed to the attacker; therefore, the proposed technique is safe. As the communication process has a one-sided transfer type, it is not a challenge-response structure.

In the security assessment [4, 5], we assess the security of the proposed protocol by the verification of satisfying security requirements using automated validation of internet security protocols and applications (AVISPA) as a formal verification tool. AVISPA assesses the security by deriving possible security threats. Figure 3 shows the verification result. As a result, in the SUMMARY, SAFE is displayed; this means that the proposed protocol is safe.

4   Conclusions

A designated PC service has been adopted to restrict the use of terminals by security threats of identity verification methods supported from the existing Internet banking service. Nevertheless, the designated PC service applied to a terminal using the service did not verify the security assessment and did not define the evaluation criteria; therefore, the service was exposed to security threats. To address this problem, the transaction linkage technique was proposed such that linkage code is generated by combining transaction information with secret information, but this technique was also exposed to various security threats. For these reasons, the current designated PC service and transaction linkage technique do not ensure security; therefore, we proposed a transaction linkage technique combining a designated terminal to solve these problems. The technique proposed in this paper can counteract and analyze all security threats; thus, the online identity verification method is also improved. We consider that the safety of Internet banking services can be enhanced by applying the proposed protocol.

Acknowledgements. This research was supported by Basic Science Research Program through the National Research Foundation of Korea (NRF) funded by the Ministry of Education (NRF-2015R1A6A3A01019717).

References

1. Hole, K.J., Moen, V., Tjostheim, T.: Case study: Online banking security. IEEE Secur. Priv. **4** (2), 14–20 (2006)
2. Baek, J., Newmarch, J., Safavi-Naini, R., Susilo, W.: A survey of identity-based cryptography. In: Proceedings of Australian UNIX Users Group Annual Conference, pp. 95–102, September 2004
3. Zhou, J., Gollmann, D.: Evidence and non-repudiation. J. Netw. Comput. Appl. **20**(3), 267–281 (1997)
4. Haley, C., Laney, R., Moffett, J., Nuseibeh, B.: Security requirements engineering: A framework for representation and analysis. IEEE Trans. Softw. Eng. **34**(1), 133–153 (2008)
5. Mirkovic, J., Dietrich, S., Dittrich, D., Reiher, P.: Internet Denial of Service: Attack and Defense Mechanisms. Radia Perlman Computer Networking and Security. Pearson Education, London (2004)

Development of ADDA (Additional Data) Algorithm for IoT Security and Privacy

Oliver M. Junio$^{(\boxtimes)}$ and Jasmin De Castro-Niguidula

Technological Institute of the Philippines, Manila, Philippines
oliverjunio@uphsl.edu.ph, jasniguidula@yahoo.com

Abstract. Internet becomes one of the basic necessity of a person. From simple sharing of data and information, internet nowadays offers millions of things such as free storage, free communication. Privacy and Security Issues are being compromise with the so many things that Internet provided. Billions of IoT devices will be released in the market by 2020 [1]. To secure connection between devices, the researcher added additional security using ADDA Algorithm. This algorithm will add additional blocks to the traditional encryption for additional security to the gateway of a particular IoT device. There are three (3) parameters to be used in this study to provide security and privacy for the IoT sites and devices. These are the accuracy, encryption speed and decryption speed of data. In this study, the researcher explains the step-by-step details how the ADDA algorithm works and make IoT devices secured for day to day use by making a new algorithm. With the results generated, ADDA algorithm gives additional protection to already encrypted data by adding characters based on the algorithm created. The result of encrypting data using ADDA algorithm was exceptional due to high percentage rate of the test conducted.

Keywords: Internet of Things · IoT · Privacy · Security

1 Introduction

Internet offers one thousand and one data and information, publish anywhere and can be access everywhere. The rise of the internet increases cybercrime this include security and privacy issues. There are three (3) identified main factor that arises the growth of internet this includes First, the development of small scale technology, Second, the inexpensiveness of the technology, and third the presence of online storage [1].

In 2020, 24 Billion of IoT Devices will be released in the market. Along with the development of the technology, is also the growth of security issues such as (1) Public Perception, (2) Vulnerability to Hacking, (3) Readiness of Company to handle security issues and (4) which of the security provider really provides security. Another concern is the privacy issues accompanied to IoT such as (1) uncontrollable volume of data, (2) Unwanted Public Profile, (3) arise of eavesdropping and (4) consumer confidence of finding everything via Internet [2].

IoT revolutionizes how individuals and corporations interact with one another. Security and privacy issues can be resolved by means of competitive advantages network technologies has been developed over the years. Direct connections to a server

© ICST Institute for Computer Sciences, Social Informatics and Telecommunications Engineering 2018
J. J. Jung et al. (Eds.): BDTA 2017, LNICST 248, pp. 9–18, 2018.
https://doi.org/10.1007/978-3-319-98752-1_2

can be limited or track down by means of embedded electronics, a good software engineer and administrator plus a good connectivity [3].

Based on the survey done by the internet world stats usage and population statistics, Philippines ranked 15 out of 20 to the top 20 countries with the highest number of internet users having 102,624,209 population and 54,000,000 registered internet users. This means that Internet is part of Philippine community daily living. The inexpensiveness of the technology cost i.e. the production of mobile or smart phone that offers internet connections and the low cost of communication provider are some of the factors that causes the growth of internet usage [4].

With the growing number of Internet users and providers together with the information published via world wide web this paper aims to determine IoT security and privacy issues in the Philippines.

This paper is organized as follows: Sect. 1 defines the state of the internet and IoT security and privacy. Section 2 introduces related work. Issues on privacy and security is discussed in Sect. 3, Sect. 4 shows the different mechanism on how issues on privacy and security can be prevented and Sect. 5 provides conclusions. There are three (3) parameters to be used in this study to provide security and privacy for the IoT sites and device. These are the accuracy, encryption speed and decryption speed of data.

2 Related Works

The modernization of communications that offers automatic connection to internet whenever there is an access made it possible for every person. IoT offers (1) SNS (Social Networking Sites), the connection it offers from one point to another point, made a convenience way of sharing files, (2) Cloud storage, where users can access files as long as there are internet connections, (3) Search engine that can dig every simple and complex data needed by the subscribers [5].

Different means of sharing files and how IoT can be a good help to a daily endeavor a person has. Security and privacy of IoT varies from (1) how people use IoT, (2) where it is connected, (3) policy it handles, (4) security algorithms it have and (5) requirements needed to be verified before connection took place. IoT was used as a Librarian main communication with the aid of mobile technology. A sustainable connection to the internet gives the company a minimal expenses of sharing files [6].

When it comes to IoT security, The Internet of Things, an emerging global Internet-based technical architecture facilitating the exchange of goods and services in global supply chain networks has an impact on the security and privacy of the involved stakeholders. Measures ensuring the architecture's resilience to attacks, data authentication, access control and client privacy need to be established. An adequate legal framework must take the underlying technology into account and would best be established by an international legislator, which is supplemented by the private sector according to specific needs and thereby becomes easily adjustable. The contents of the respective legislation must encompass the right to information, provisions prohibiting or restricting the use of mechanisms of the Internet of Things, rules on IT-security-legislation, provisions supporting the use of mechanisms of the Internet of Things and the establishment of a task force doing research on the legal challenges of the IoT [11].

Another research study about privacy challenges from the Internet of Things, these services can be provisioned using centralized architectures, where central entities acquire, process, and provide information. Alternatively, distributed architectures, where entities at the edge of the network exchange information and collaborate with each other in a dynamic way, can also be used. In order to understand the applicability and viability of this distributed approach, it is necessary to know its advantages and disadvantages – not only in terms of features but also in terms of security and privacy challenges. The purpose of this paper is to show that the distributed approach has various challenges that need to be solved, but also various interesting properties and strengths [12].

While the general definition of the Internet of Things (IoT) is almost mature, roughly defining it as an information network connecting virtual and physical objects, there is a consistent lack of consensus around technical and regulatory solutions. There is no doubt, though, that the new paradigm will bring forward a completely new host of issues because of its deep impact on all aspects of human life. In this work, the authors outline the current technological and technical trends and their impacts on the security, privacy, and governance. The work is split into short- and long-term analysis where the former is focused on already or soon available technology, while the latter is based on vision concepts. Also, an overview of the vision of the European Commission on this topic will be provided [13].

Describe developments towards the Internet of Things (IoT) and discuss architecture visions for the IoT. Our emphasis is to analyze the known and new threats for the security, privacy and trust (SPT) at different levels of architecture. Our strong view is that the IoT will be an important part of the global huge ICT infrastructure ("future Internet") humanity will be strongly relying on in the future with relatively few data centers connected to trillions of sensors and other "things" over gateways, various access networks and a global network connecting them. While the infrastructure is globally connected, it is divided into millions of management domains, such as homes, smart cities, power grids, access points and networks, data centers, etc. It will evolve both bottom-up and top-down. An important question is what consequences a bottom-up and top-down construction of the IoT infrastructure has for the security, privacy and trust and what kind of regulation is appropriate [14].

Embedded, mobile, and cyberphysical systems are ubiquitous and used in many applications, from industrial control systems, modern vehicles, to critical infrastructure. Current trends and initiatives, such as "Industrie 4.0" and Internet of Things (IoT), promise innovative business models and novel user experiences through strong connectivity and effective use of next generation of embedded devices. These systems generate, process, and exchange vast amounts of security-critical and privacy-sensitive data, which makes them attractive targets of attacks. Cyberattacks on IoT systems are very critical since they may cause physical damage and even threaten human lives. The complexity of these systems and the potential impact of cyberattacks bring upon new threats [15].

The Internet of Things consists of various platforms and devices with different capabilities, and each system will need security solutions depending on its characteristics. There is a demand for security solutions that are able to support multi-profile platforms and provide equivalent security levels for various device interactions. In addition, user privacy will become more important in the IoT environment because a lot of personal information will be delivered and shared among connected things. Therefore, we need mechanisms to protect personal data and monitor their flow from things to the cloud. In this talk, we describe threats and concerns for security and privacy arising from IoT services, and introduce approaches to solve these security and privacy issues in the industrial field [16].

IoT introduces the usage of technology to both businesses and consumers. The adaptation of technology as part of people daily lives becomes part of the commodity needed by the society. The solutions it offers and the security mechanism injected on it are sometimes neglected by the consumers, for them as long as technology made their lives easier is more than enough [8]

To ensure IoT Security, Fuzzy logic is best to determine the protocols and algorithms included in the selected research sites with respect to its reliability and efficiency in providing security and privacy [9].

One of the research studies shows how IoT Security was implemented in the network layers and how the algorithm was used to provide efficient security mechanism. Protocols such as RSA and EAS are the major protocols used within their selected sites along with the encryption and decryption algorithm fused together with the protocol [17].

With the aid of IoT, Burt (2016) contrast Hahn (2017) on his believes in IoT. In his report to the United States National Security, Burt pointed out that IoT is a big disguise that uses technology as front end and served as a spy back door. This manner of hiding the true identity of provider and subscribers to the public comprises the security and often result to identity theft and eavesdropping problem. Monteiro (2015) uses Fuzzy logic to provide results for reliability and efficiency in checking the security and privacy to IoT device data which is the same in this study with the help of ADDA algorithm.

3   Methodology

Rapid Application Development was used in creating ADDA algorithm using Visual Studio 6.0.

The researcher used experimental approach to obtain the result of accuracy, encryption speed and decryption speed of data inputted into the IoT device.

To secure connection between devices, the researcher added additional security using ADDA Algorithm. This algorithm will add additional blocks to the traditional encryption for additional security to the gateway of a particular IoT device.

Figure 1 shows how the proposed method constitutes of encrypted data. The ADDA Algorithm will get the encrypted data and will add additional block and pattern that will add confusion and diffusion to possible attack.

Encryption Algorithm

Step 1: Get the encrypted data.
Step 2: For every 4 bits of block add additional block.
Step 3: Add New character to the additional block.

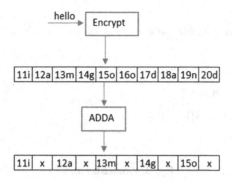

Fig. 1. ADDA algorithm

Step 4: Place each data to each equivalent container.
Step 5. Save the new encrypted data.

Figure 2 shows the ADDA algorithm program. It composes of command buttons (open file, copy to source folder, 1st Encryption, 2nd encryption (ADDA), decryption and exit system.

Fig. 2. Adda algorithm main program

4   Results and Discussion

To visualized and see how ADDA algorithm works, following figures were presented.

Encrypting Text

Figure 3 shows original text file named source.txt which will be encrypted later.

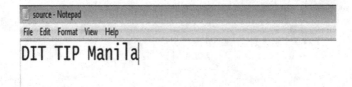

Fig. 3. Source.txt file

Figure 4 shows the encrypted data (1st encryption command button). The 1st encryption will encrypt the source.txt file by converting characters including spaces into hexadecimal code and in between there is a special character inserted. The source. txt file will be replaced by source_adda_omj.txt.

Fig. 4. Encrypted text (1st encryption) source_adda_omj.txt file

Figure 5 shows the encrypted data (2nd encryption ADDA command button) 0 using ADDA algorithm which every character in Fig. 2 was converted to binary plus in every 4 blocks a randomized special character was being inserted.

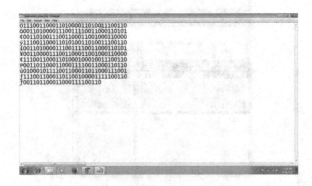

Fig. 5. Destination_adda_omj.txt file

Figure 6 shows the decrypted data (decryption command button) which brings back the original text and filename (source.txt)

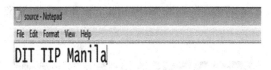

Fig. 6. The original text and file name source.txt

Figure 7 shows the encryption speed and decryption speed in milliseconds results of data that was encrypted and decrypted. Hardware specification where the program was run is intel core i7 with 8gb RAM running in windows 7 64 bit operating system. It also shows that the decryption speed in most tests conducted was doubled as compared to the encryption speed.

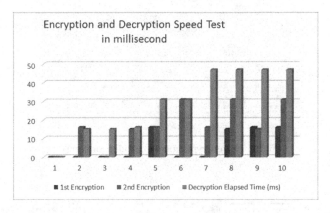

Fig. 7. Speed test report (text file)

Encrypting Image
Figure 8 is the original image that will be encrypted later using ADDA algorithm.

Fig. 8. Original image to be process

Figure 9 is the 1st encryption of the file source.jpg. it took 115752 ms to complete the encryption

Fig. 9. 1st encryption (source_adda_omj.jpg)

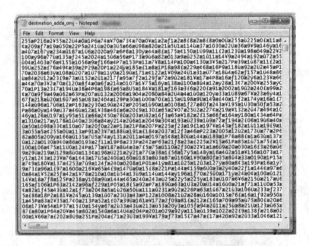

Fig. 10. 2nd encryption (destination_adda_omj.jpg)

Figure 11 shows the decrypted image which is exactly the same of the original image.

Table 1 shows the accuracy of image resolution before and after encryption and decryption occurs. The accuracy of the system to bring back the original file in its original resolution is 100%.

Figure 10 shows the encrypted file using ADDA algorithm

Fig. 11. The original image after decryption

Table 1. Accuracy test of image resolution

	Original File Size	Original resolution in pixel	File Size after decryption	Resolution after decryption	Accuracy of resolution after
1	84.4	512 × 384	475	512 × 384	100%
2	84.4	512 × 384	475	512 × 384	100%
3	84.4	512 × 384	475	512 × 384	100%
4	84.4	512 × 384	475	512 × 384	100%
5	84.4	512 × 384	475	512 × 384	100%
6	84.4	512 × 384	475	512 × 384	100%
7	84.4	512 × 384	475	512 × 384	100%
8	84.4	512 × 384	475	512 × 384	100%
9	84.4	512 × 384	475	512 × 384	100%
10	84.4	512 × 384	475	512 × 384	100%

5 Conclusion

Security and privacy are one the major concern of IoT users, with the aid of ADDA algorithm additional security will be added to the traditional encryption. Privacy were protected using two procedures (1) by adding blocks to the original blocks and (2) by randomly inserting special characters. Based on the result, the encrypted data will be more secured and can be used privately since blocks of data are encrypted with the help of ADDA algorithm. The accuracy result of encrypting data using ADDA algorithm was exceptional due to high percentage rate of the test conducted.

References

Smith, M.: Protecting privacy in an IOT – connected world. Inf. Manag. **49**, 36–39 (2015)

Meola, A.: How the Internet of Things will affect security and privacy (2016). http://www.businessinsider.com/internet-of-things-security-privacy-2016-8

Navetta et al.: The Security, Privacy, and Legal Implications of the Internet of Things (IoT) Part One – The Context and use of IoT (2015). http://www.dataprotectionreport.com/2015/05/the-security-privacy-legal-implications-of-the-internet-of-things-iot-part-one-the-context-and-use-of-iot/

de Argaez, E.: Miniwatts Marketing Group, International Division of Miniwatts de Colombia Ltda, Carrera 7, Bogota, Colombia (2017)

Bian, J., et al.: Mining Twitter to assess the public perception on the Internet of Things. PLoS ONE **11**, 1–14 (2016)

Hahn, J.: The Internet of Things: mobile technology and location services in the libraries. Libr. Technol. Rep. **53**(1), 1–28 (2017)

Burt, J.: IoT Could be used by Spies, U.S. Intelligence Chief Says. eWeek, p. 1, 2 December 2016

Patra et al.: Securing IoT devices and gateways (2016). http://www.ibm.com/developerworks/library/iot-trs-secure-iot-solutions1/index.html

Monteiro, E., et al.: Security for the Internet of Things: a survey of existing protocol and open research issues. IEEE Commun. Surv. Tutor. **17**(3), 1294–1312 (2015)

Suo, H.: Security in the Internet of Things: A Review. Guangdong Jidian Polytechnic Guangzhou, China (2012). https://www.researchgate.net/publication/254029342_Security_in_the_Internet_of_Things_A_Review

Weber, R.H.: Internet of Things-new security and privacy challenges. Comput. Law Secur. Rev. **26**(1), 23–30 (2010)

Roman, R., Zhou, J., Lopez, J.: On the features and challenges of security and privacy in distributed Internet of Things. Comput. Netw. **57**(10), 2266–2279 (2013)

Medaglia, C.M., Serbanati, A.: An overview of privacy and security issues in the Internet of Things. In: Giusto, D., Iera, A., Morabito, G., Atzori, L. (eds.) The Internet of Things, pp. 389–395. Springer, New York (2010). https://doi.org/10.1007/978-1-4419-1674-7_38

Kozlov, D., Veijalainen, J., Ali, Y.: Security and privacy threats in IoT architectures. In: Proceedings of the 7th International Conference on Body Area Networks, pp. 256–262. ICST (Institute for Computer Sciences, Social- Informatics and Telecommunications Engineering), February 2012

Sadeghi, A.R., Wachsmann, C., Waidner, M.: Security and privacy challenges in industrial Internet of Things. In: 2015 52nd ACM/EDAC/IEEE Design Automation Conference (DAC), pp. 1–6. IEEE, June 2015

Hwang, Y.H.: IoT security & privacy: threats and challenges. In: Proceedings of the 1st ACM Workshop on IoT Privacy, Trust, and Security, p. 1. ACM, April 2015

Liu, Y., Zhou, G.: Key technologies and applications of Internet of Things. In: 2012 Fifth International Conference on Intelligent Computation Technology and Automation (ICICTA), pp. 197–200. IEEE, January 2012

Fine-Grained Data Traffic Management System Based on VPN Technology

Yiming Liu[1], Zhen Cheng[1], Ziyu Wang[2], Shihan Chen[1],
and Xin Su[1(✉)]

[1] College of IOT Engineering, Hohai University, Changzhou 213022, China
15995086362@163.com, lte_5g@yeah.net,
1299068532@qq.com, leosu8622@163.com
[2] NanJing Ivtime, U2-1205 NO. 12 East Mozhou Road,
Town of Future Network, Nanjing 210000, China
wangziyu@ivtime.com

Abstract. With the development of the wireless network and mobile Internet service, the bandwidth of wireless network has increased obviously, and the number of wireless network user and data traffic generated by user terminal are steadily on the increase. Development in different economic and social sectors not only require higher network speed but also need low-cost information networks. The market needs a complete set of solutions, which can provide a comprehensive management system for the operation and billing of fine-grained data traffic. In this paper, we focus on the management system of fine-grained data traffic operation, which is support reverse charge and realize the fine-grained data traffic. Therefore, the proposed method can save user's cost and improve economic efficiency.

Keywords: Fine-grained data traffic · VPN technology
Data flow identification

1 Introduction

With the development of the wireless network and mobile Internet service, the bandwidth of the wireless network have increased obviously, and the number of wireless network user and data traffic generated by user terminal are steadily on the increase.

Development in different economic and social sectors not only require higher network speed but also need low-cost information networks. At present, there has been a contentious issue that users should pay operators when they use mobile phones to surf the Internet. For example, a transportation company called "Tencent" runs a bus called "WeChat" carrying nearly nine hundred million Internet users to "mobile internet", and who will pay the tolls? It is obvious that the bus driver should pay the tolls. However, the current situation cannot meet fine-grained data traffic management. The market needs a complete set of solutions, which can provide a comprehensive management system for the operation and billing of fine-grained data traffic [1–3].

© ICST Institute for Computer Sciences, Social Informatics and Telecommunications Engineering 2018
J. J. Jung et al. (Eds.): BDTA 2017, LNICST 248, pp. 19–25, 2018.
https://doi.org/10.1007/978-3-319-98752-1_3

In order to provide users with better experiences and eliminate their expense concerns, we hope to implement a detailed and fine-grained billing of data traffic according to the types of business or user's custom. The research target of this paper is aimed at developing a fine-grained data traffic management system which can provide a detailed and fine-grained billing of data traffic. The user's various applications will be processed uniformly like data collection, statistical analysis, data mining, safety management, and other processing, to achieve a goal of detailed and fine-grained billing according to the types of applications, business or specific data traffic packages [4].

2 Analysis on Data Traffics

Through China-mobile Communication Corporation in Changzhou, we obtained the monthly average data traffic of 4 base stations in Hohai University (in the first half of 2016). E-UTRAN Cell Global Identifier (ECGI) of four base stations respectively is 4600-871713-12 (31.8193° N, 119.97166° E), 4600-872639-1 (31.82152° N, 119.97886° E), 4600-341532-1 (31.81869444° N, 119.9787778° E), 4600-341533-1 (31.8206° N, 119.975° E). Figure 1 shows the exact location and monthly data traffic of the four base stations, and the difference between the monthly data traffic of 4600-341533-1 and 4600-341532-1 is significant. After analysis, the former is close to the student dormitory and the traffic users are mostly students. Students are the majority of the current traffic users and live 24 h a day on campus. The latter is close to office buildings and classrooms, and part of traffic users are teachers, while they use traffics mainly during the day.

Fig. 1. Four base stations at Hohai University of Changzhou Campus.

At the same time, B2C mode is asymmetric, charge between companies and consumers cannot reverse. However, companies want to provide a better user experience, for example, let users under any circumstances to be able to shop or play games online, etc. Therefore, Alibaba and some other companies are willing to cooperate with the three operators to pay the fee to consumers and attract consumers to buy their products. However, many small and medium Internet enterprises are not willing to reverse the charge, because the present marketing system, management analysis system, network management, billing system of all operators does not support the reverse charge. Only upgrade system can achieve this goal but they are not willing to bear the upgrade cost. Because of this, the reverse charge plan is difficult to conduct.

3 Analysis of Existing VPN Technologies

Traffic transmitted through a tunnel, such as a VPN or protocol agency. In this paper, we use the mature, controllable and secure VPN technology. Currently, there are several common VPN technologies [3–6]:

3.1 Point to Point Tunneling Protocol (PPTP)VPN

The Point to Point Tunneling Protocol (PPTP) is a new enhanced security protocol developed based on the Point to Point Protocol (PPP). The protocol supports VPN, and can enhance security through Password Authentication Protocol (PAP), Extensible Authentication Protocol (EAP), etc. The PPTP can create, maintain and terminate a tunnel through connection control, and can encapsulate PPP frames by using the Generic Routing Encapsulation (GRE). Before encapsulation, the PPP frames' payload, that is effective transmission data, need a mixed process of encryption and compression first, and then make the remote users directly connect the Internet or other network by accessing the Internet Service Provider (ISP). The PPTP VPN is developed by Microsoft and is standardized. The GRE tunnel is created dynamically with TCP controlling tunnel (plaintext), and the GRE tunnel encapsulating real user data traffic—the PPP payload. The encryption of payload is merely based on a self-contained encryption protocol—MPPE, the head of GRE is all clear, and security level is not high. Moreover, other details are not transparent.

3.2 Layer 2 Tunneling Protocol (L2TP) VPN

L2TP is a kind of Virtual Private dial-up Network (VPDN) tunnel protocol. VPDN refers to accessing the public network using the dial-up function of public networks (such as ISDN or PSTN), and realize a virtual private network which can provide access services for enterprises, small Internet service provider (ISP) and mobile workforce, etc. VPDN can provide an economical and effective point-to-point connection between remote users and private enterprise networks. L2TP is an industry-standardized Internet tunnel protocol, which is similar to the PPTP protocol, for example, both of them require encryption of network data flow. The difference is that, for example, PPTP requires Internet Protocol (IP) network, while L2TP requires packet-oriented point-to-point connection. PPTP uses a single tunnel and L2TP USES multiple tunnels; L2TP provides the header compression, tunnel verification, while PPTP does not support.

3.3   Internet Protocol Security (IPsec) VPN

IPSec works like packet-filtering firewall, which can be viewed as an extension of packet-filtering firewall. When an IP packet is received, the packet-filtering firewall uses its header to match in a rule table. When a matching rule is found, the received IP packets will be processed by the packet-filtering firewall in accordance with the method decided by the rule table. There are only two processing works here: discarding or forwarding. The IPSec decides the process of IP packets received by searching the Security Policy Database (SPD). But unlike the packet-filtering firewall, IPSec handles the IP packet with the IPSec process except discarding, directly forwarding (bypassing the IPSec). It is because the new process that more network security than packet-filtering firewall can be provided.

3.4   Secure Sockets Layer (SSL) VPN

SSL is encrypted between the fourth and fifth layers and is often certified with digital certificates. Hyper Text Transfer Protocol over Secure Socket Layer (HTTPS) is a type of client less SSL VPN, and it is vulnerable to man-in-the-middle. Currently, apple IOS 10. X does not support PPTP VPN, while the high version of android only supports IPsec VPN and L2TPVPN. Therefore, considering the compatibility and safety of the scheme, this paper mainly adopts IPsec VPN to realize traffic fine-grain operation management system. At the same time, for security, we use the Internet Key Exchange version 2 (IKEv2) protocol allows server better preventing Disk Operation System (DOS) attack in terms of the secret key exchange control. And the building of ISAKMP SA in Phase 1 only need 4 message exchanges instead of 6. IKEv2 is used to add a notify payload, and whenever there is a client connection to IKE server, the server responds a notify payload containing a cookie. Meanwhile, the client receives the notify message and contains the cookie in the next connection request. The connection will be established if the cookie is verified legal, otherwise, it will be viewed as a DOS attack, and dose not establish a connection. The server side simply saves a cookie that occupied a few bytes of memory and can be released quickly. IKEv2 is used in authentication, negotiate encryption, hash algorithm, the encryption of data traffic and the establishment of IKE SA, IP SEC SA. Security is ensured by the certification (Pre-shared password, digital certificate), Advanced Encryption Standard (AES), Triple Data Encryption Standard (3DES), Secure Hash Algorithm (SHA), Message Digest Algorithm (MD5), Anti-replay window.

The main software part proposed in this paper and its description are illustrated in Table 1.

4   System Design Objectives and Requirements

In this article, the flow fine-grain operation management system constructed based on the principle of standardization, will strictly follow the requirements of the relevant technical standards and business to practice overall planning and unified construction arrangement; In the meantime, based on the principle of openness, the system follows

Table 1. Software components and its descriptions of the system

Software components	Descriptions
APP client	Data flow identification and distribution It is suggested to provide the channel download business application and provide the starting entrance of the APP
SDK (Android version)	Android version SDK can provide the fine-grained data separating capacity and can realize APP packaging call
Data forwarding server (tunnel forwarding, security encryption)	Forwarding the data flow in server Making the security encryption and forwarding

an open architecture and adopts an open interface protocol and development platform to provide users with a unified and open capability call. In addition, business maintenance and development are not dependent on equipment manufacturers to ensure the continuous upgrading and development of the business. In terms of security, our system will be designed strictly according to the application of the telecommunications level. The system software and hardware architecture should fully consider the security strategy and mechanism of the whole system operation. In view of the security requirements of various business processes, various security technologies are adopted to provide users with perfect security. Finally, a software design framework with mature and stable operation instance is adopted [2, 4, 5].

5 System Architectures

The type of applications installed in users' equipment will be identified automatically when the user starts the data flow workshop SDK or APP. The network configuration information will automatically synchronize with data traffic management gateway. Data traffic management gateway will send feedback to users based on the current traffic load of network and the information. When the user adds a certain application to the directional flow packet, the flow workshop SDK or APP will create the pre-defined corresponding VPN tunnel to the traffic load device according to the type of applications. Data traffic packets are generated by a specific APP, and local SDK or APP can identify the packet by identifying the APP ID of the application started by users. And then according to the relevant protocol of VPN tunnel and the group package way of flow workshop SDK or APP to re-group packages, send to the specified tunnel, the flow is unpacked at the end of the tunnel, and then to the business server of SP. The data reverse process is similar.

Figure 2 shows the flow of VPN-call. It is initiated by the APP to register the login authentication, and the policy server manages the terminal users and data service nodes to perform uniform user access configuration. The APP internally make traffic identification according to the product used, and identifies the dynamic IP address of the current SP to bind with APPID temporarily. The APP server launch VPN resource request to the policy server, and make the policy server choose the nearest data server to access based on the IP address of the APP itself. Meanwhile, policy configuration will be forwarded to data load server based on the dynamic IP address of SP recognized by the APP. When the resource is ready, the server request APP to return the confirmation message and SA. The APP user uses the data server information and SA to log in the specified data load server, establish the virtual traffic tunnel for the identified APP data traffic. These four steps complete the establishment of VPN, and the identification and forwarding of data traffic [3–6].

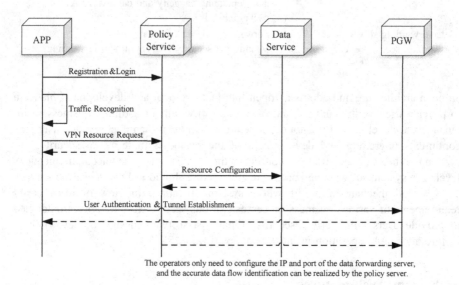

The operators only need to configure the IP and port of the data forwarding server, and the accurate data flow identification can be realized by the policy server.

Fig. 2. The flow of VPN-call in typical business scenario.

The operators only need to configure the IP and port of the data forwarding server, and the accurate data flow identification can be realized by the policy server. Operators' billing management platform can obtain traffic statistics information through API docking. The way of billing can also be developed by the policy server itself.

6   Conclusions

With the development of the wireless network and mobile Internet service, the bandwidth of the wireless network have increased obviously, and the number of the wireless network user and the data traffic generated by user terminal are steadily on the increase. Development in different economic and social sectors not only require higher network speed but also need low-cost information networks. The market needs a complete set of solutions, which can provide a comprehensive management system for the operation and billing of fine-grained data traffic. In this paper, we focus on the management system of fine-grained data traffic operation, which is support reverse charge, and can realize the fine-grained division of data traffic. We bring a new type of consumption model for customers to save money based on traffic data reverse charge for customers. We use VPN to encapsulate applications separately that will help operators easily calculate the flow data for each application and reversely charge from online enterprises. Many online enterprises will be willing to cooperate with operators to pay the fee to consumers because they want to provide a better user experience, for example, let users under any circumstances to be able to shop or play games online, etc., and attract consumers to buy their products. We have experimentally employed the proposed reverse charge system, and the questionnaires have shown that this is a win-win model for both customers and online enterprises. We hope the proposed system will be finally implemented in practice.

Acknowledgements. This research is supported by the Natural Science Foundation of Jiangsu Province under Grant BK20160287.

References

1. Wan, H., Lin, Y., Zhihao, W., Huang, H.: Discovering typed communities in mobile social networks. J. Comput. Sci. Technol. **27**(3), 480–491 (2012)
2. Hou, J., Xu, B., Xu, L., Wang, D., Xu, J.: A testing method for web services composition based on data-flow. Wuhan Univ. J. Nat. Sci. **13**(4), 455–460 (2008)
3. Yan, T., Wang, B.: Grid architecture model of network centric warfare. J. Syst. Eng. Electron. **17**(1), 121–125 (2006)
4. Chen, J., Chen, Z., Wang, Q., Fang, Y.: Spatial database management system of China geological survey extent. J. China Univ. Geosci. **14**(3), 250 (2003)
5. Xiao, Y., Chen, Y.: Effcient distributed skyline queries for mobile applications. J. Comput. Sci. Technol. **25**(03), 523–536 (2010)
6. Jianming, Y., Guoxin, W., Junming, W.: Rebuilding the extrant's architecture with VPN. J. Southeast Univ. (English Edition), (1), 69–74 (1999)

Image Processing

Segmentation of Hanacaraka Characters Using Double Projection Profile and Hough Transform

Liliana Liliana$^{(\boxtimes)}$, Singgih Mardianto Soephomo,
Gregorius Satia Budhi, and Rudy Adipranata

Informatics Department, Petra Christian University, Surabaya, Indonesia
{lilian, greg, rudya}@petra.ac.id

Abstract. In doing segmentation of Hanacaraka character, Javanese ancient character, one of Indonesian's ethnic ancient character in Java island, the difficulties that occur is the inconsistency of the space between lines, the size of the character and the thickness. Inconsistencies between row spacing and letter size are caused by the letters of the pair, the last vowel and consonant letters in one phoneme. While the thickness is inconsistent due to the writing style of the Hanacaraka itself.

Image Preprocessing needs to be done to get input without skew. To improve skewed text documents, we used Hough transforms to predict the edges of the text area. After that, to segment the line and then continue with segmentation of each character, horizontal projection profile is used and then proceed with vertical.

The result of this segmentation method is good for printed documents. Segmentation process of handwriting documents has difficulty because each row in the document is uneven and very tight between the rows. Those matters cause them overlap. When the line segmented wrongly, the entire character on the line will be not segmented as well. This problem can be eliminate using connectivity test. Before this, it need to segment the line with the overlap area. The character part of below or above the main character can be eliminate because it is not connected to the main character.

Keywords: Segmentation · Hanacaraka character · Projection profile
Hough transform · Image processing

1 Introduction

A culture basically has a wide variety of variations. In general, the various types of culture are dances, songs, local games; local languages etc. one variation of cultures which is also very visible and often used is the local language. According to a source from Kompas 2012, in continuing study, which was taking samples at 70 sites in Maluku and Papua, the number of languages and sub languages across Indonesia reached 546 languages [1].

Along with the development of technology and global communication, the condition of our culture has been increasingly eroded. Hundreds of local languages in

© ICST Institute for Computer Sciences, Social Informatics and Telecommunications Engineering 2018
J. J. Jung et al. (Eds.): BDTA 2017, LNICST 248, pp. 29–37, 2018.
https://doi.org/10.1007/978-3-319-98752-1_4

Indonesia are threatened to extinct. There is an estimation of 746 local languages in Indonesia, yet only more than 400 languages and sub languages has been successfully mapped [2].

Letters in local languages are knows as a form of writing or a representation of that local language. One of the languages having special letters as a form of writing of that language is Javanese with Javanese writing or better known as Javanese characters. The Javanese letters, also known as Hanacaraka and Carakan, is one of the Indonesian traditional characters, used to write Javanese language. In daily lives, the use of Javanese characters is generally replaced with Latin letters which were first introduced by the Dutch in 19th century [3].

Nowadays, there have been enough efforts to preserve the Javanese letters, either by the government or professional circle. One of the good efforts is the development of android-based Hanacaraka application teaching Javanese characters. According to Tekkomdik, besides developing the application, the digitalization of cultural contents such as puppets, macapat songs and also documentary video, would also be launched [4].

2 Theories

2.1 Hough Transform

The data input, in the form of captured image from a digital camera, has the tendency to slant or skew. The skew found is all deviation of the image causing the result after the process of inputting using the hardware differs from the initial image or shape [5]. To solve this problem, Hough Transform is the method used to detect the skew at the image [5–8].

Hough Transform is a technique of edge linking and boundary detection, commonly used in image processing [5–8]. The purpose of this method is to find the shape of the object in a class of objects using the voting procedure. This voting procedure is conducted in a parameter from the object candidate obtained as local maxima. This parameter will later be called accumulator, specifically formed in the algorithm to calculate Hough Transform.

Figure 1 is representing the geometric interpretation from parameter θ and ρ. A horizontal line has $\theta = 0°$, with ρ having positive value. A vertical line also has $\theta = 90$, with ρ having positive value at intercept y or $\theta = -90$ with ρ having negative value at intercept y.

The unique calculation concept from Hough Transform is the grouping of parameters ρ and θ into an *accumulator array*. The distance expected at that parameter is $-90 \leq \theta \leq 90$ and $-D \leq \rho \leq D$, where D is the maximum distance between opposite ends in an image. The following steps are to calculate Hough Transform:

(1) Perform a Looping for all pixels at the input image. For every non-background pixel P_{ij}.
(2) Perform the looping from–D up to D. The mathematical Eq. 1 to calculate the value of D is as follows:

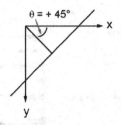

Fig. 1. Representation of a line.

$$D = \sqrt{((\text{image\_height})^2 + (\text{image\_width})^2)} \tag{1}$$

(3) Calculate the value of ρ for each angle of $-90 \le \theta_i \le 90$.
(4) Do the rounding for the value of ρ using the mathematical Eq. 2.

$$\rho = x \, \cos(\theta) + y \, \sin(\theta) \tag{2}$$

(5) Do addition at Hough Matrix H_{ij}.

2.2 Segmentation Based on Projection Profile

Projection Profile is a histogram consisting of the number of foreground pixels accumulated along the parallel line in a document [9–13]. In several other cases, Projection Profile was used to skew estimation, text line segmentation, page layout segmentation, etc. [9]. The implementation at this application is by dividing the Projection Profile into two types. They are horizontal projection profile and vertical projection profile. The horizontal projection profile is used to find the line region from the document, whereas the vertical projection profile is used to take the character out of each line.

Below is the mathematical equation for horizontal projection profile and vertical projection profile:

$$HPP(y) = \sum_{l \le x \le n} F(x, y) \tag{3}$$

$$VPP(x) = \sum_{l \le y \le m} F(x, y) \tag{4}$$

The samples of the horizontal projection profile and vertical projection profile images can be viewed at Figs. 2 and 3.

3 Analysis

The main problem in the segmentation of Javanese characters will be resolved in this study is the skew of the documents as the input, the italics and the overlapping writings between lines due to the inconsistency of line spacing or characters sticking together, see Fig. 4. The skew document is a problem that often occurs in DCR

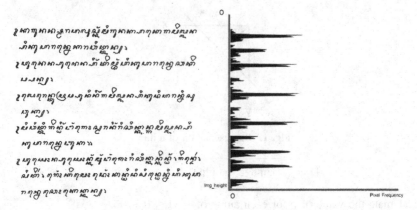

Fig. 2. The input of digital image and its result of horizontal projection profile

Fig. 3. An image cut of the first row from Fig. 2, and its vertical projection profile

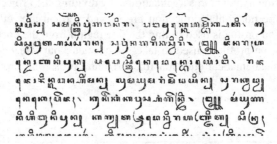

Fig. 4. The sample of an image with overlapping characters in lines

(Digital Character Recognition) application [5–8]. The cause of the problem is the error that occurs during inputting. The skew that occurs is generally $-15° \leq x \leq 15°$.

It has been found that there have been many overlapping or sticking writings condition in a document with Javanese characters. Italics and overlapping writings have brought on the result of the segmentation less optimal. The Overlapping writings in this document were found horizontally (overlap writings between columns) and vertically (overlap in lines). This problem occurs because characters such *as vowel* (special character put above the main character), and *sandhangan* (special character added below the main character) and *carakan* (special adding character for phoneme which adopted from foreign language). Meanwhile the slanted writings are found due to the writing style of the Javanese characters itself. The slanted writing style is not something unusual, as currently some normal texts have slanting writings form which are often called italic. Italics are usually found in a handwritten document.

Proposed Method

Figure 5 shows the system we developed. Some pre-processing need to be performed to get a non-skewed binary image as input for the segmentation process. From "bitmap to array" until "binary thresholding process" are the pre-processing. After get a binary image, the first step is repair a skewed input. Projection profile cannot be performed on a skewed image. To detect and correct the skewed document, Hough transform is used because it has better performance than scanline method [5]. This method is will determine the border of text area [5–8].

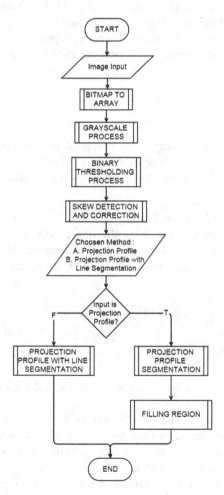

Fig. 5. Flowchart of the system

If the input has inconsistency space between raw, then the system will run segmentation process without line segmentation. To improve the quality from segmentation without line segment, it will perform filling region procedure. This procedure tries to reconstruct the missing part after the segmentation.

The other problem is overlapped writings. To solve this problem, there have been several studies conducted. The former study on several kinds of character, such as Kannada [6, 11], Devanagari [7], Arabic [8], Urdu [9], Gurmukhi [10], Chinese character [12] and Oriya [13]. These study ware using projection profile and connected regions methods to do the segmentation of the characters in the writings. However, in several cases, projection profile in the segmentation of Javanese characters cannot be fully applied. The structure of the writings and the unique characteristics of Javanese characters can make projection profile fail.

Kumar repairs the corrupt character with water reservoir technique to get a good result [10], Mamatha uses morphological operation [11], Tripathy uses line segmentation specifically to detect the writing where the line of the writing could not be found using the projection profile, as the writings were overlapped or touched the lines under. Our system uses double projection process to improve the quality of character segmentation.

4 Experiment

The testing performed was to compare the output of the program against the manual calculation. In this testing, the sample of data were classified into two groups, data of the inconsistent spaces between lines (or overlap rows) and data of the inconsistent size and type of characters.

The Inconsistent Space Between Lines
At the sample data as seen at Fig. 6, as the first process of projection profile was conducted, only 43 out of 558 characters could be segmented. At the second process of projection profile, 96 out of 558 characters could be segmented.

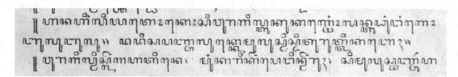

Fig. 6. Photograph of script with line inconsistency [14]

The Inconsistent Sizes and Types of Characters
The example of sample data taken from the photograph belonged to the sample data having differences in sizes and types of characters. At the sample data as seen at Fig. 7, the writings with different sizes and types could not be segmented as the lines at different parts of the writings were overlapped with other writings. The average result of the testing showed that 15% of the writings could be segmented. Table 1 shows the result of segmentation using data in Fig. 7.

Fig. 7. Photograph of script with size inconsistency [15]

Table 1. Testing using data in Fig. 7

No	Line on the sample	Writing	Line output		Writing output		%
			Right	Wrong	Right	Wrong	
1	25	814	8	17	254	560	31
2	22	558	1	21	79	479	14
3	12	249	12	0	246	3	98
4	14	396	7	3	222	174	56
5	23	805	0	23	0	805	0
6	30	1560	1	29	55	1505	3
7	28	1537	1	27	28	1509	1
8	26	1375	4	22	208	1167	15

Using sample data with certain condition, inconsistency space between rows, different font size such as shown in Fig. 7 and different thickness as shown in Fig. 8. Usually, when the hand stroke upper, the line will thinner than when the hand moves lower. Some handwritten documents or even printed documents will have this writing style. On a printed document, this kind of style will not lead to failed segmentation process because every stroke separated well.

For some document like shown in Fig. 7, cannot segmented because some lines of the document laid in differently with the main part.

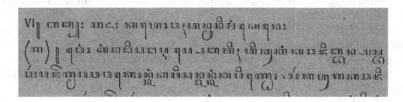

Fig. 8. Hanacaraka writing style [16]

5 Conclusion

This system performs better with good hanacaraka image. Printed text images have a constant space between row or not enough space to separate rows, consistent font size and thickness. Those kinds of input will yield good result. The result shown in Table 1 is come from problematic input. Some of them have inconsistency space between rows, and some other have different font size or different thickness in one single character. Overlap rows will lead to fail line segmentation. This will affect the character segmentation also.

Based on the testing result, the projection profile method, on average can perform the segmentation of the writing at a document by 22% for the group of photograph data having inconsistent spaces between lines. 77% for the group of photograph data having consistent spaces between lines.

Filling region procedure can help the segmentation to reach 63.5% of character overlap segmentation

The other problem come from skewed document can be resolve if the skew less than 95°. Document with consistent space between row can be segmented over 75%. Different thickness can be solved as long as each character separately well.

Acknowledgment. This research was funded by DIPA Directorate General of Research and Development Reinforcement (Direktorat Jenderal Penguatan Riset dan Pengembangan) no. SP DIPA-042.06.1.401516/2017, fiscal year 2017.

References

1. Akuntono, I.: Nasional – Kompas, 1 September 2012. Kompas Cyber Media. http://nasional. kompas.com/read/2012/09/01/12030360/Mau.Tahu.Jumlah.Ragam. Bahasa.di.Indonesia
2. Sudirman, M.: Republika, 5 Maret 2013. http://www.republika.co.id/berita/koran/news-update/14/03/04/n1wzn0-bahasa-daerah-semakin-punah
3. Agfa Monotype Corporation: Monotype, 1 January 2000. http://www.monotype.co.uk/ NonLatin/wt_info/info_javanese.html
4. Nbi: Hanacaraka, Aplikasi Android untuk Belajar Aksara Jawa, 20 Desember 2013. Tribunne News Cyber Media. http://jogja.tribunnews.com/2013/12/20/hanacaraka-aplikasi-android-untuk-belajar-aksara-jawa/
5. Kishan, A.C., Sharda, V.: Skew Detection & Correction in Scanned Document Images. Department of Computer Science and Engineering, National Institute of Technology Rourkela, Orissa (2009)
6. Ramappa, M.H., Srikantamurthy, K.: Skew detection, correction and segmentation of handwritten Kannada document. Int. J. Adv. Sci. Technol. **48**, 71–88 (2012)
7. Garg, R., Garg, N.K.: An algorithm for text line segmentation in handwritten skewed and overlapped Devanagari script. Int. J. Emerg. Technol. Adv. Eng. **4**(5), 114–118 (2014)
8. Alginahi, Y.M.: A survey on Arabic character segmentation. Int. J. Doc. Anal. Recogn. **16** (2), 105–126 (2013)
9. Javed, M., Naghabushan, P., Chaudhuri, B.: Extraction of Projection Profile, Run-Histogram and Entropy Feature Straight from Run-Length Compressed Text Documents. Department of Studies in Computer Science, University of Mysore, Kolkata (2014)

10. Kumar, M., Jindal, M.K., Sharma, R.K.: Segmentation of isolated and touching characters in offline handwritten Gurmukhi script recognition. Int. J. Inform. Technol. Comput. Sci. **2**, 58–63 (2014)
11. Mamatha, H.R., Srikantamurthy, K.: Morphological operations and projection profiles based segmentation of handwritten Kannada document. Int. J. Appl. Inform. Syst. (IJAIS) **4**(5), 13–19 (2012)
12. Mei, Y., Wang, X., Wang, J.: A Chinese character segmentation algorithm for complicated printed documents. Int. J. Sig. Process. Image Process. Pattern Recogn. **6**(3), 91–100 (2013)
13. Tripathy, N., Pal, U.: Handwriting segmentation of unconstrained Oriya text. Sadhana **31**(6), 755–769 (2006)
14. BPAD Tentara Pelajar: Dongeng Koetjing Setiwelan. Yogyakarta, p. 5 (1992)
15. BPAD Tentara Pelajar: Langendriya, Yogyakarta, p. 7 (1938)
16. Rijksblad: Yogyakarta No 1, p. 9 (1936)

Face Expression Extraction Using Eigenfaces, Fisherfaces and Local Binary Pattern Histogram Towards Predictive Analysis of Students Emotion in Programming

Anna Liza A. Ramos[1], Jerico M. Flor[2], Michael M. Casabuena[2],
and Jheanel E. Estrada[2(✉)]

[1] Saint Michael's College of Laguna, Biñan, Philippines
annakingramos@yahoo.com.ph
[2] Technological Institute of the Philippines, Manila, Philippines
flor.jerico.m@gmail.com, mmcasabuena@firstasia.edu.ph,
jheanelestrada29@gmail.com

Abstract. Emotion plays an important role to assess individual reaction and responses depends on the degree of encounter, scenario and experience. In this paper, the study examines the emotions of the five (5) randomly selected students according to the Basic Emotions and Non-Basic Emotions while performing their Java Programming activity task. The study utilized RaspberryPi and Smartphone to capture an image, OpenCV application to analyze the image and application of algorithms: the Fisherfaces, Local Binary Pattern Histogram and Eigenfaces to determine the most like emotion. In this study, the Fisherfaces algorithm showed the highest average accuracy rate of 47.93% among the algorithms. Specifically marked an emotion of "happy and surprise" with accuracy rate of 100% which means that the students perform the activity with knowledge and skills. This result can be used by the experts to consider the emotion as part of assessment hence it may also serve as a tool for effective decision making.

Keywords: Facial expression · Emotion · Predictive analysis

1 Introduction

Emotions plays a crucial role in our lives it should be managed and interpreted accurately. In fact, this has been the concern of scientific inquiry in psychology since emotion is consider as the driver in decision making [1] and the source of moral judgement [2]. This may be interpersonal, intrapersonal, socio and cultural functions [3] thus, constitute to individual level, dyadic level, group level and cultural level [4]. Moreover, emotions made a great influence on how people manage emotion, understand emotion, using emotion and perceiving emotion which is evidently noticed in the academe and in workplace environment [5]. These factors lead the affective computing focused area.

© ICST Institute for Computer Sciences, Social Informatics and Telecommunications Engineering 2018
J. J. Jung et al. (Eds.): BDTA 2017, LNICST 248, pp. 38–46, 2018.
https://doi.org/10.1007/978-3-319-98752-1_5

Affective Computing, on the other hand has been a great interest in intelligent interactive and pattern recognition, emotion expression learning and understanding of emotion [6]. Due to the contributory impact of technologies there are innovative solutions which are being developed in the form of audio, visual and text that are capable to recognize expressions, interpret gestures and simulate appearance to determine emotion. Many companies integrate analysis of emotions and sentiments to enhance customer relationship management and recommendation system [7] however, there are some issues and concerns since emotions [8] is broad because it is more than thoughts, it also depends on the response of the body and it is difficult to achieve its accuracy and progress towards cognitive modelling because of various modalities to be considered. But, this would not hinder the discovery and contribution of affective computing in producing good results that may affects areas of concern.

1.1 Related Works

In the advancement of technology, the emotion can be detected and analyzed based on the facial expression, in fact facial expression is used to determine the current emotion pertaining to stress [9], used to capture individual identity for security purposes to recognize malicious intention through gestures and surveillance, access control to a building [10], used in virtual meetings to assess person reactions, [11] used by deaf people to convey a message, assess a customer feedback on a certain product, evaluation of student emotions and behavior in computer programming using Python Language which resulted to confusion, frustration and boredom [12], used to assess the emotion of participants interact with different computer interfaces activities resulted to consider the non-basic basic emotion such as engaged, bored, frustrated and confusion be part of Affective Computing research [13].

Several Studies applied various approaches to detect emotions through Facial Expression to achieve accuracy and performance through different various models, algorithms and applications. Notably, Neural Network Model using JAFEE [14], Curvlet Transform which is fast, less complex and less redundant and Online Sequential Extreme Learning Machine (OSELM) with radial basis function (RBF) which increase classification performance to 95.17% recognition rate [15], Automated Learning Free Facial Landmark Detection Technique which performs in different resolutions and accurate for classification of the Six universal expression [16] and SMQT features, split up SNoW classifier to detect face using standard pattern and Principal Component Analysis in terms of luminance, chrominance to locate the eyes based on valley points [17], Kanada-Lucas Tomasi Tracker which is accurately used for face detection based on distance from the camera, brightness and contrast and the Skin color pixel value that ranges from 120 to 180 pixel in value and Tree Naive Bayes Classifier [18], the face detection that employed Active Appearance Model to locate landmarks [19] used the Support Vector Machine classification using nearest neighbor rule and Extended Cohn-Kanade (CK+) datasets thus resulted to 87.7% performance based on the local and global features of the face. In Eigenfaces and Fisherfaces face recognition shows better recognition accuracy of 97.50% and 95.45% based Euclidean Distance using Bray Curtis [20]. For complexity issues eigenfaces is applied with Gaussian Curvature to detect 3D image [21] and it is the best algorithm to extract feature of a face with the application of Fisher Linear Discriminant and its classify using Dynamic Fuzzy Neural Networks to reduce errors [22]. Fisherfaces on the other hand is used to treat image with eyeglasses based on genetic algorithm [23] and this also used for gender recognition with fuzzy iterative self-organizing technique with accuracy rate 95.55% [24] the FGGA System on Chip applied the Local Binary Pattern Histogram recorded an accuracy rate of 79.33% [25].

2   Method

In this part, the study applied a systematic approach to examine the emotions portrayed by the respondents.

Figure 1 showed the methodology on how the study is being conducted and facilitated. In this framework there are two types of approaches applied by the researchers, capture a video then convert into frames and capture an image. These images are stored in the database assigned in the study annotated by the experts and with incorporation of the CK/C+ database which will be used to match the real-time video and analyzed the data based on the employed algorithm in the study which is expected to produce certain emotions.

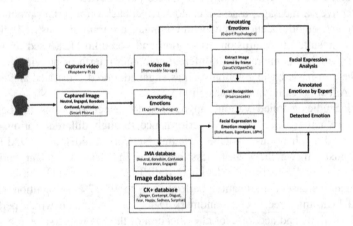

Fig. 1. Methodology framework

2.1   Data Gathering

The study applied the Random Sampling in selecting the sample population to be used in the training set and in the testing part which aim to capture certain emotion. The participants are represented from Colleges and Universities who offered Information Technology Program.

In the training set part, the sample population is twenty (20) student participants composed of 10 males and 10 females from First Asia Institute of Technology while in the testing part, the sample population is five (5) students composed of 2 males and 3 females from Technological Institute of the Philippines (TIP).

2.2   Tools

Raspberry PI

In this experiment that researchers used Raspberry Pi 3 model V to provide high resolution and can capable to capture the image in 300 × 300 pixel in size from the video sequence and used smart phone LG G3 13-megapixel F/2.4 29 mms to capture an image portrayed by the students with the presence of the experts in the field. OpenCV also used for face detector that process 90–95% of clear images [14].

Figure 2 showed the images of the student who portrayed the Non-basic emotions annotated the expert in the field.

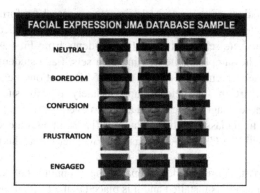

Fig. 2. Facial expression data sets

Databases: JMA Database & Cohn-Kanade Database (CK+)

The JMA database is referred to a database where all the captured images gathered by researchers will served as the training data sets and this will be incorporated and as additional pattern of emotion such as neutral, engaged, frustration, confusion, boredom and Cohn-Kanade Database (CK+) L.

Table 1 showed the number of images used in the study. These images portrayed different types of emotion with its corresponding numbers of images.

Table 1. CK/CK+ Database

Coding	Emotion	No. of images
1	Anger	45
2	Contempt	18
3	Disgust	60
4	Fear	25
5	Happy	69
6	Sadness	28
7	Surprise	83

To execute and populate the database, the captured images are manually classified according to the five (5) emotions (neutral, boredom, confusion, frustration and engaged) then normalizing the training set using OpenCV for face detection and face images are cropped into 300×300 pixel in size and integrate to the CK/CK+ database that have the seven (7) basic emotions (anger, contempt, disgust, fear, happy, sadness and surprise). These datasets are trained through the application of algorithms Fisherfaces, Local Binary Pattern Histogram and Eigenfaces to further improved the datasets being developed.

2.3 Data Gathering Procedure

To facilitate the conduct of the data gathering, the researchers formalized the study through a formal letter of request indicating the requirements needed in the study. In this study, the researchers selected twenty (20) students from First Asia Institute of

Technology for the sample of the training data sets. Each student are requested to portray five (5) pose per emotion with a total of 25 emotions per student with the guidance of an expert in the field to appropriately portray such emotion using Smartphone, then these images are cropped and displayed for re-evaluation purposes of an expert before its final classification of emotion. These images are stored in the JMA database together with the CK/CK+ database that will served as the training sets used in the study.

For the testing part, the researchers prepared the Computer Laboratory for the data gathering, the RaspberryPi is installed and it is placed at the back of the monitor and the camera is attached at the center top head area of the monitor to capture the image proportionately. The identified students are requested to position themselves in front of the monitor and advised to answer the provided machine problem in 15 min in a continuous video recording. The raw video file are stored and converted into frames with an interval of 5 min to determine the emotions portrayed by the students. This frame of images are subject for processing.

2.4 Analysis

Presented the system architecture of the Facial Expression Extraction Prediction of Student Emotion in Programming (FEEPSEP).

Figure 3 showed the real-time video is used to extract the images into frames and then be analyzed and classified using the Haar-based Cascade Classifier to serve as the input images. These images are then compared to the JMA and CK/CK+ databases which is already been trained to detect the most like emotions according to the respective results of the three algorithms.

Eigenfaces is implemented since it efficiently processes the time and storage with accuracy rate of 90% with the Principal Component Analysis [16], the application of Local Binary Pattern Histogram which performed very well in terms of texture classification and segmentation, image retrieval, surface inspection and [32] showed enough discriminated faces and non-faces faster. And, the Fisherfaces algorithm which is good distortion analysis of the faces such as illumination, facial expression and pose

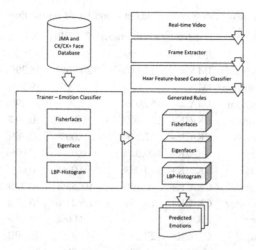

Fig. 3. FEEPSEP system architecture

variations of the face. This three (3) features of algorithm applied in this study will surely made a significant effect on its performance to detect certain emotion.

The images has been analyzed using percentage formula:

$$\frac{\text{Total image detected per algorithm}}{\text{The total number of detected emotion of the (3) algorithms}} \text{ multiplied by 100}$$

3 Results and Discussion

In this study, the researchers developed a prototype to detect certain emotion. In result, the solution showed that the Fisherfaces algorithm have the highest accuracy of 47.93% among the algorithms used in the study. Specifically, the Fisherfaces algorithm achieved 100% accuracy presented in Table 2. "Happy and Surprise" means that the majority of the students in the study perform the programming task provided during the testing part.

Table 2 Showed the comparative results of the algorithm based on combine JMA and CK/CK+ database, the Fisherfaces marked an accuracy rate of 47.93%, LBPH is 26.09% accuracy rate and the Eigenfaces marked the lowest accuracy of 25.16%. However, looking at the results of the JMA and CK/CK+ database, the CK/CK+ database significantly produce a remarkable highest accuracy results of 100% observed in Fisherfaces Algorithm, 98.53% in LBPH Algorithm and 82.76% percent in Eigenfaces Algorithm than the JMA Database annotated by the experts only produced the highest accuracy rate of 69.66% in Fisherfaces Algorithm, 38.82% in LBPH Algorithm and 82.76% in Eigenfaces Algorithm.

Table 2. Comparative results of detected emotion based on the three (3) algorithms

No.	Emotion	Database	%Fisherfaces	% LBPH	% Eigenfaces
11	Engage	JMA	23.19%	38.82%	37.98%
10	Frustration	JMA	24.31%	20.30%	55.39%
9	Confusion	JMA	69.66%	10.59%	19.75%
8	Boredom	JMA	50.03%	17.99%	31.99%
7	Surprise	CK/CK+	100.00%	0.00%	0.00%
6	Sadness	CK/CK+	17.24%	0.00%	82.76%
5	Happy	CK/CK+	100.00%	0.00%	0.00%
4	Fear	CK/CK+	0.00%	0.00%	0.00%
3	Disgust	CK/CK+	0.74%	98.53%	0.74%
2	Contempt	CK/CK+	18.19%	65.27%	16.54%
1	Anger	CK/CK+	96.72%	0.00%	3.28%
0	Neutral	CK/CK+	27.18%	44.43%	28.39%
		Total	47.93%	26.90%	25.16%

4 Conclusion and Recommendation

Based on results, the Fisherfaces algorithm is more flexible and efficient considering some issues on the integration of the JMA database because the annotation of the expert alone may possibly have encountered inaccuracy in identifying real emotions.

In addition, by looking at the results of 47.93% accuracy rate, there are some factors also be considered, the lack of training dataset presented in Table 1 and in the testing part showed in Table 2, showed 0% detection rate that may significantly create a negative effect on the accuracy result, therefore the study highly recommend the number of training data sets is consistent, the integration of the new training data set should have further analysis, used solutions that will assist the expert to re-validate the results and may apply all approaches of the three algorithms if the new training data set is embedded to the existing database like the CK\CK+ database which is highly tested and validated to ensure higher accuracy rate in detecting emotions.

5 Acknowledgement

The researchers extend our thanks to the TIP Management for the support to conduct the research, To the students of First Asia Institute of Technology who served as face model in the study and the students of Technological Institute of Technology – Manila for participating the programming activity and also Mr. Alfrie Sarmiento who served as sample model. To the experts from Saint Michael's College of Laguna who validate and annotate the images captured namely Ms. Ruth Ann Musngi, Ms. Marjorie Magno and Ms. Julie Quintao for their great contribution to make this study successful. And, to the panel members who gave their great ideas, contribution to make this study more significant and relevant namely, Dr. Jasmine Niguidula, Dr. Alexander Hernandez and Dr. Enrico Chavez.

References

1. Jaiswal, S., Bhadauria, S.S., Jadon, R.S.: Comparison between face recognition algorithm-Eigenfaces, fisherfaces and elastic bund graph matching. J. Glob. Res. Comput. Sci. **2**(7), 187–193 (2011)
2. Huebner, B., Dwyer, S., Hauser, M.: The role of emotion in moral psychology. Trends Cogn. Sci. **13**(1), 1–6 (2008)
3. Hwang, H., Matsumuto, D.: Functions of Emotion. San Francisco State University (2016)
4. Keltner, D., Haidt, J.: Social Functions of Emotion at Four Level of Analysis. University of California-Berkeley, USA & University of Virginia, Charlottesville, VA, USA (1999)
5. Bracket, M.A., Rivers, S.E., Salovey, P.: Emotional intelligence: implications for personal, social, academic and workplace success. Soc. Pers. Psychol. Compass **5**(1), 88–103 (2011). Yale University
6. Niedental, P.M., Brauer, M.: Social functionality of human emotion. Annu. Rev. Psychol. **63**, 259–285 (2013). University of Winconsin-Madison
7. Cambria, E.: Affective Computing and Sentiment Analysis. Nanyang Technological University (2016)
8. Picard, RW.: Affective Computing Challenges. MIT Media Laboratory, Room E15-020G Ames Street, Cambridge, MA 02139-4307, USA (2003)
9. Greene, S., Thapliyal, H., Caban-Holt, A.: A survey of affective computing for stress detection. IEEE Consum. Electron. Mag. **5**(4), 44–56 (2016)
10. Choi, J., Lee, S.C., Yi, J.: A Real Time Face Recognition using Multiple Mean Faces and Dual Mode Fisherfaces. School of Electrical and Computer Engineering, Sungkyunkwan University. IEEE (2001)
11. Inam, H., Malik, A., Hayat, M., Asraf, A.: A Survey on Facial Expression Recognition Technology and its Use in Virtual System. Department of Software Engineering, Fatima Jinnah Women University, The Mall Rawalpindi 46000, Pakista. MECS Publisher (2015)
12. Bosch, N., Mello, S.D., Mills, C.: What emotions do novices experience during their first computer programming learning session. In: International Conference on Artificial Intelligence in Education (2013)
13. D'Mello, S., Calva, R.A.: Beyond the Basic Emotions: What Should Affective Computing Compute (2013)
14. Enami, S., Suciu, V.P.: Facial recognition using OpenCV. J. Mob. Embed. Distrib. Syst. **4** (1), 38–43 (2012)
15. Ucar, A, Demir, Y., Guzelis, C.: A new facial expression recognition on curvelet transform and online sequential extreme learning machine initialized with spherical clustering. Neural Comput. Appl. **27**(1), 131–142. Springer, Heidelberg (2016). https://doi.org/10.1007/s00521-014-1569-1
16. Happy, S.L., Routray, A.: Automatic facial expression recognition using features of salient facial patches. IEEE Trans. Affect. Comput. **6**(1), 1–12 (2015)
17. Halder, R., Sengupta, S., Pal, A. Ghosh, S., Kundu, D.: Real time facial emotion recognition based on image processing & machine learning. Int. J. Comput. Appl. **139** (2016)
18. Chinchamalatpure, A., Jain, A.: Survey on human face expression recognition. Int. J. Comput. Sci. Mobile Comput. **3**(10), 20–24 (2014)
19. Hsu, F.S., Lin W.Y., Tsai, T.W.: Automatic Facial Expression Recognition for Affective Based on Bag of Distances. Department of Computer Science and Information Engineering. National Chung Cheng University (2013)

20. Shyam, R., Singh, Y.N.: Evaluation of Eigenfaces and Fisherfaces using Bray Curtis dissimilarity metric. Department of Computer Science & Engineering, Institute of Engineering and Technology, India (2012)
21. Kurihara, T., Shu, Z., Ono, N., Ando, S.: A facial authentication system using complex-valued Eigenfaces. SICE Annual Conference in Sapporo, Hokkaido Institute of Technology, Japan (2004)
22. Tangquan, Q., Huiwen, D., Weipong, H.: Face recognition using Eigenfaces-Fisher linear discriminant and dynamic fuzzy neutral network. IEEE (2010)
23. Song, C.F., Yin, B.C., Sun, Y.F., Liu, L.: Eyeglasses Fisherfaces Base Glasses-Face Recognition, Multimedia and Intelligence Software Technology Beijing, Municipal Key Laboratory, Beijing University of Technology, Beijing, China, Eight International Conference on Machine Learning and Cybernetics (2009)
24. Yijun, D., Xiabo, L., Wujun, C., Qianzhou, X.: Gender recognition using Fisherfaces and a fuzzy iterative self-organizing technique. In: 10th International Conference on Fuzzy System and Knowledge Discovery (2013)
25. Stekas, N., Heuvel, D.V.: Face recognition using local binary patterns histograms (LBPH) on a FPGA-based system on chip. In: IEEE International Parallel and Distributed Processing Symposiums Workshops (2013)

Context Awareness

Improving Indoor Positioning Performance in an Emergency Deployment System

JaeMin Hong[1,2], KyuJin Kim[1,2(✉)], and ChongGun Kim[1,2]

[1] Department of Computer Engineering, Yeungnam University,
Gyeongsan, South Korea
{hjm4606,kayjay6t}@naver.com, cgkim@yu.ac.kr
[2] Department of Nursing, Kyungpook National University, Daegu, South Korea

Abstract. The proposed real time emergency position control system can treat real-time emergency messages between servers and mobile clients based on reply from the client by using multiple communication methods. Especially in the general hospital environment, in order to avoid patient anxiety caused by emergency situations, the system informs calmly the emergency situation to the person who has to react the situation by using one-way broadcast communications. The accuracy and performance of positioning system are important to increase reliability on the proposed system. The power consumption rate of mobile devices have analyzed and the process of positioning data is verified for one-way communication and two-way communications.

Keywords: Bluetooth LE · Wi-Fi · Power consumption · Indoor positioning
Fingerprint · Kalman filter

1 Introduction

In a general hospital environment, in case of emergency, it is needed to send the emergency information in real time to the persons in charge by using SMS or telephone call for respond the situation. The persons in charge with the corresponding situation have to respond as soon as possible. Based on the general emergency call method, if the emergency information cannot be timely conveyed to the persons in charge, there will delayed the deployment of the persons in charge and cannot respond be in time. A system that can send emergency information to persons in charge by using multiple communication methods when an emergency situation occurs and the persons in charge can respond to the emergency in real time is proposed. The system is also can record and manage all emergency activity information.

Figure 1 shows the multiple communication concept of the proposed system.

2 Previous and Related Studies for Indoor Positioning

In this study, the management server uses one-way broadcast communication to send emergency alarm message to persons in charge, according to the emergency reaction protocol, then change to two-way communications by confirming from the client to

© ICST Institute for Computer Sciences, Social Informatics and Telecommunications Engineering 2018
J. J. Jung et al. (Eds.): BDTA 2017, LNICST 248, pp. 49–59, 2018.
https://doi.org/10.1007/978-3-319-98752-1_6

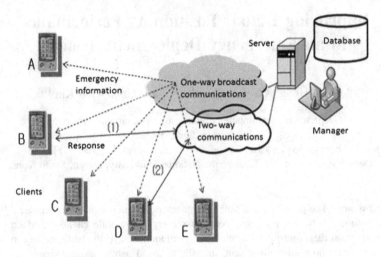

Fig. 1. The communication concept of the emergency deployment system

exchange emergency information with clients, all the activities and position of the responder are tracked in real time.

Bluetooth LE (Bluetooth Low Energy), visible light communication technology, SMS message can be used as one-way broadcast communication method, Wi-Fi and cellular mobile communications can be used as two-way TCP communication methods in this system.

2.1 Bluetooth LE

By growing importance of IoT (Internet of Things), Bluetooth LE application is also diffused. Compared with the Bluetooth, it does not need to maintain a long connection, therefore power consumption is also greatly reduced.

BLE does not need pairing that can send data to multi clients in one-way, this process is called Advertising. All the terminals which Bluetooth enabled can receive the advertising data. When a message is transmitted via advertising, the message can be receiving by multi users in the same time.

The BLE device for advertising is called beacon and it is possible to indoor positioning by using RSSI (Received Signal Strength Indication) value of beacon signal [1]. For indoor positioning using beacons, Cell ID method is mainly used. It is a method of judging in which cell the position of its own is located based on the information received from the fixed node. This method is easier and simpler than other methods, but has the disadvantage of low accuracy. Indoor positioning using beacons is mainly used for hospital, department store and etc.

2.2 Two-Way Communication and Indoor Positioning Based on Wi-Fi

Socket communications have two major transport layer protocols, TCP (Transfer Control Protocol) and UDP (User Datagram Protocol).

The UDP communication has no handshaking dialogues among the processes for session, and cannot provide any reliability control so there is no guarantee of delivery, ordering, or duplicate protection of datagrams. The TCP communication provides reliable, ordered and error-checked delivery of a stream of octets between processes which are running on hosts based on IP network. In this study, TCP is used for socket communications.

A TCP communication is requested by using the IP addresses and port numbers of two processes. The client starts the connection and if the server does not accept the response, the connection will be failed. If connection is created, data can be exchanged through the socket between the server and client. The TCP can provide error control and flow control, if the data not received properly, it can be requested again [2].

GPS (Global Positioning System) is the most widely used outdoor positioning system. But in the indoor environment, GPS cannot work properly. Therefore, many research of indoor positioning are in progress. For example, visible light communications [3], BLE signal strength [1, 9], Wi-Fi signal strength and so on can be used [4, 5]. For finding the coordinate of mobile object in indoor environment, fingerprint and triangulation methods are widely used.

The triangulation method is a geometric method of computing a coordinate by calculating each distance from three reference points. It is necessary to be able to obtain the correct distance from the wireless AP as reference point so that the correct position can be traced [10].

Fingerprint method is RSSI (Received Signal Strength Indication)-based, but it simply relies on the previously recorded data of the signal strength from several reference access points in the proper range. Storing this information in a database along with the known coordinates of the tracking device is clone in an offline phase. During the online position decision phase, the current RSSI vector at an unknown location is compared to those stored in the fingerprint DB and the closest matching position is returned as the estimated user location [8].

Wi-Fi or BLE signal strength can be used for calculating distance for using the fingerprint method.

2.3 Emergency Deployment System in a General Hospital

When a patient meet in an emergency situation, this system can send the emergency notice to the persons in charge on various departments, then by the confirming to the notice which means the corresponding departments can give an initial solution. Indoor positioning is used to guide the persons in charge to the current location of the patient, then lead those on the most appropriate route to the operating room.

In an emergency situations, excessive noise in a hospital has been attributed to negative clinical outcomes to patients and perhaps give negative performance and stress as well to all staffs. It is essential to use a plain language emergency code only for an appropriate individuals to initiate an immediate and appropriate response.

When initiating an emergency code call by the control center in a general hospital setting, the notification process for specific doctors and nurses will be initiated with single-way multiple task communication. Once the emergency code message has been effectively sent to emergency responders, staff will press button to open bilateral one to one communication and exchange information about the target locations and specific emergency operational plan with server. Start with confirmation response from emergency response staffs, initiating protocol information will be recorded in server.

Server with location tracking system manage staff's current position, moving route and the time required after confirmation response in real time. If delayed response is occurred, server sends urgent messages and emergency operational plan to emergency response staff. But if emergency response staff cannot response emergency call, server initiates a contingency plan. It provides alternate medical staff and notifies other emergency response staff about current situation. Emergency code should be finished when situation has been managed or resolved.

Every process such as the time required from beginning to end, staff location per time, situation status and so on is recorded in database. With these data, further analysis and improvement of the system will be managed.

3   Comparison of Energy Consumption on Wi-Fi and BLE

Different communication methods not only have difference on data rate and communication distance, power consumptions are also different. Therefore, the power consumption tendencies are seriously considered in the proposed system.

3.1   Required Energy for Operating Hardware Modules

The WL18xxMOD WiLink 8 Single band Combo Module developed by TI which is widely used in portable mobile devices has be selected for power consumption analysis of Wi-Fi device. One important selection reason is the chip on the device uses the same antenna for Wi-Fi and Bluetooth communications.

The chip's operating power of Wi-Fi communication showed on Table 1 [6]. It can be seen that, depending on the negotiation data rate and the encoding method, the operating power of chip at 49 mA to 85 mA when the chip used as a Wi-Fi receiver. As a Wi-Fi transmitter, the operating power of chip surged to 238 mA to 420 mA.

The chip's operating powers of Bluetooth BR (Basic Rate) and EDR (Enhanced Data Rate) communication showed on Table 2 [6], The operating power of each case in a range from 178 μA to 33 mA.

Table 3 showed the chip's operating powers of Bluetooth LE (BLE). The current values are from 124 μA to 266 μA [6].

The power consumption of Wi-Fi is a thousand times than that of Bluetooth LE.

3.2   Energy Consumptions Based on Software Structure

The power consumption of TCP/IP based two-way communication protocol via Wi-Fi, one-way Bluetooth LE communications, and using Google Cloud Message (GCM) Push [7] are tested. The confirmation of power consumption software used in this study is done based on the BroadcastReceiver function of the Android operating system. Figure 2 is the flow chart of this program. When an event is transmitted from the Android system, the receiver application which is corresponded can be executed.

Table 1. Operating current of Wi-Fi communication

	SPECIFICATION ITEMS	TYP (AVG)	UNITS
Receiver	Low-power mode (LPM) 2.4 GHz RX SISO20 single chain	49	mA
	2.4 GHz RX search SISO20	54	
	2.4 GHz RX search MIMO20	74	
	2.4 GHz RX search SISO40	59	
	2.4 GHz RX 20 M SISO 11 CCK	56	
	2.4 GHz RX 20 M SISO 6 OFDM	61	
	2.4 GHz RX 20 M SISO MCS7	65	
	2.4 GHz RX 20 M MRC 1 DSSS	74	
	2.4 GHz RX 20 M MRC 6 OFDM	81	
	2.4 GHz RX 20 M MRC 54 OFDM	85	
	2.4 GHz RX 40 M MCS7	77	
Transmitter	2.4 GHz TX 20 M SISO 6OFDM 15.4 dBm	285	
	2.4 GHz TX 20 M SISO 11 CCK 15.4 dBm	273	
	2.4 GHz TX 20 M SISO 54 ODFM 12.7 dBm	247	
	2.4 GHz TX 20 M SISO MCS7 11.2 dBm	238	
	2.4 GHz TX 20 M MIMO MCS15 11.2 dBm	420	
	2.4 GHz TX 40 M SISO MCS7 8.2 dBm	243	

Table 2. Operating currents of Bluetooth

USE CASE	TYP	UNIT
BR voice HV3+sniff	11.6	mA
EDR voice 2-EV3 no retransmission+sniff	5.9	mA
Sniff 1 attempt 1.28 s	178.0	µA
EDR A2DP EDR2(master). SBC high quality – 345 kbps	10.4	mA
EDR A2DP EDR2(master). MP3 high quality – 192 kbps	7.5	mA
Full throughput ACL RX: RX-2DH5	18.0	mA
Full throughput BR ACL TX: TX-DH5	50.0	mA
Full throughput EDR ACL TX: TX-2DH5	33.0	mA
Page scan or inquiry scan (scan interval is 1.28 s or 11.25 ms, respectively)	253.0	µA
Page scan and inquiry scan (scan interval is 1.28 s and 2.56 s, respectively)	332.0	µA
* Current is measured at output power as follows: BR at 11.7 dBm; EDR at 7.2 dBm		

In power consumption evaluation, data package sends total 60 times in 10 min. Then the power consumption results are measured after completion of communications. Samsung Galaxy S3 is used as receiver, ipTime A3004 wireless router is used as Wi-Fi transmitter, Raspberry Pi 3 is used as BLE transmitter in this experiment. The results of power consumption experiment are shown on Table 4.

Table 3. Operating current of Bluetooth LE(BLE)

USE CASE	TYP	UNIT
Advertising, not connectable	131	µA
Advertising, discoverable	143	
Scanning	266	
Connected, master role, 1.28 s connect interval	124	
Connected, slave role, 1.28 s connect interval	132	
*All current measured at output power of 7.0 dBm		

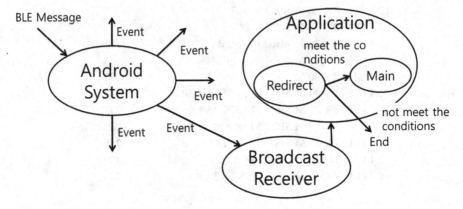

Fig. 2. Flow chart of power consumption analysis program

Table 4. Result of Power consumption experiment

Communication method	TCP/IP based own Communication protocol via Wi-Fi	Bluetooth LE	GCM Push
Battery consumption	9%	2%	1%

As shown at the tables, Bluetooth LE communication and GCM Push consumes less power than Wi-Fi communication in both tests.

It is definite that the needed operating power based on hardware modules of Wi-Fi is a thousand times higher than those of Bluetooth LE communications. From a software point of view, the difference of power consumption is also obvious. Therefore, in this system, GCM Push and BLE communication are preferred for sending an emergency one-way notification to conserve power, and Wi-Fi communications is used for ending the specific solution among the server and clients because a large amount of data exchange is required in this process.

3.3 The Process for Switching Between One-Way and Two-Way Communications

BLE requires less energy but, Wi-Fi requires a relatively large amount of energy. For the robustness of the device which should always wait for communication, BLE is used for one-way communication. And Wi-Fi is used for two-way communication for indoor positioning and emergency information transmission.

Figure 3 shows the process for switching between one-way and two-way communications.

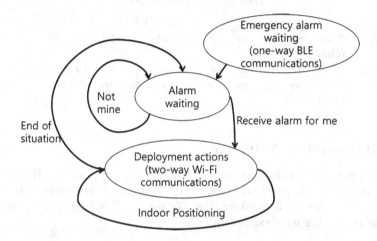

Fig. 3. The process for switching between one-way and two-way communication

4 Indoor Positioning Based on Fingerprint

For indoor positioning using fingerprint, The tracking data must be collected in advance. We store the pre-collected tracking data based on RSSI value at various indoor positions in the database.

Figure 3 shows an experimental environment with 4 APs in corridor. Tracking data is collected at 108 points.

4.1 Methods of Collecting Tracking Data

For accurate tracking data, accurate RSSI values should be collected. But RSSI values have errors and are measured differently each time. Mean, median, Kalman filter can be used to collect accurate RSSI value.

The mean is the average value of all data. The operation is simple, but if there is extremely large error, the value can be distorted.

The median is the middle value among the collected data sorted by size. If the number of data is odd, select the middle value, and if it is even, average the two numbers on both sides of the middle.

The Kalman filter is based on measurements made according to time and can give more accurate results than other measurements [11]. It can process data containing noise, and can optimal statistical prediction.

Table 5 shows RSSI values of AP1 measured at any point, and Fig. 4 shows comparison of results of mean, median, and Kalman filter.

Table 5. RSSI values of AP1

	1	2	3	4	5	6	7	8	9	10
RSSI (dBm)	-64	-58	-59	-59	-64	-57	-56	-56	-58	-59
	11	12	13	14	15	16	17	18	19	20
RSSI (dBm)	-60	-58	-59	-59	-62	-57	-56	-56	-58	-64

4.2 Calculating Similarity Distance

Fingerprint uses not the absolute distance but the similarity distance. The similarity distance should be calculated using the RSSI value at each test point with the pre-collected RSSI value in the tracking data. We use the K-NN (K-Nearest Neighbor) algorithm to calculate the similarity distance.

$$D_i = \left(\sum_{j=1}^{n} |S_j - S_{ij}|^q \right)^{1/q} \tag{1}$$

In K-NN algorithm of Formula (1), 'n' is the total number of wireless APs, 'j' is the ID number of wireless APs, and 'i' is the test point number where the signal strength is collected. 'q' is a distance constant, that has 1 for Manhattan distance method and 2 for Euclidean distance method. 'D_i' is the similarity distance, 'S_j' is the signal strength of the jth AP at the positioning point, and 'S_{ij}' is the signal strength of the jth AP at the ith collection point [12].

The nearest point can be found by comparing the similarity distance obtained at each point. But it is inefficient to calculate the similarity distance of all 108 points and then compare them.

After finding the nearest AP using the Cell ID method, it is possible to reduce the amount of calculation by comparing only the data around the AP. Figure 5 shows the range where the signal intensity of each AP is the strongest, and we can select from 18 to 36 points for each AP.

When calculating using the selected data, the amount of computation is reduced to 1/3 compared to the total data, and the computation time also decreases.

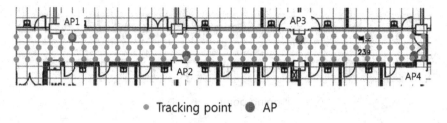

Fig. 4. An experimental environment in corridor

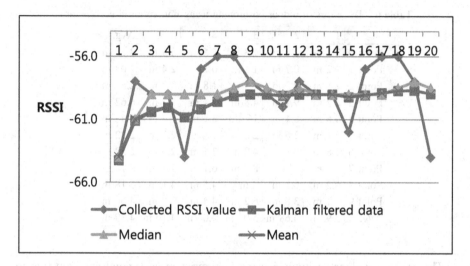

Fig. 5. Comparison results of Mean, Median and Kalman filter

4.3 Position Tracking and Error Analysis

Tracking data is collected by filtering 30 RSSI values at each tracking point. And then, as shown in Fig. 6, position tracking was performed 5 times at 10 positioning points (Fig. 7).

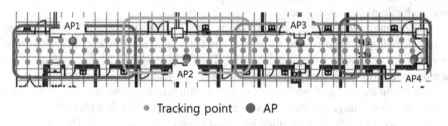

Fig. 6. The range where the signal intensity of each AP is the strongest

The errors in each positioning point are as shown in Table 6.

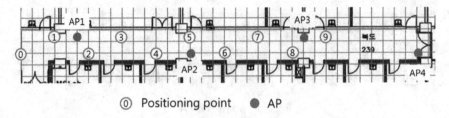

(0) Positioning point   ● AP

Fig. 7. 10 positioning point in experimental environment

Table 6. The errors of indoor positioning using fingerprint in each point

	1	2	3	4	5	Avg.
Point 0	3.1 m	2.6 m	1.8 m	5.1 m	3.3 m	3.18 m
Point 1	1.2 m	0.7 m	1.1 m	0 m	2.4 m	1.08 m
Point 2	2.1 m	2.2 m	2.5 m	1.8 m	3.3 m	2.38 m
Point 3	5.7 m	9.1 m	4.8 m	6.4 m	7.1 m	6.62 m
Point 4	4.5 m	8.1 m	16.2 m	8.4 m	5.4 m	8.52 m
Point 5	3.1 m	2.6 m	2.2 m	1 m	2.2 m	2.22 m
Point 6	1.8 m	0 m	4.8 m	3.1 m	2.9 m	2.52 m
Point 7	5.4 m	11.6 m	8.2 m	6.1 m	7.4 m	7.74 m
Point 8	8.2 m	6.1 m	3 m	4.5 m	4.1 m	5.18 m
Point 9	4.1 m	3.8 m	6.2 m	4.4 m	3.8 m	4.46 m
Avg.	3.92 m	4.68 m	5.08 m	4.08 m	4.19 m	4.39 m

The errors of indoor positioning using fingerprint in each point are from 0 m to 16.2 m. The average error is about 4.39 m. For the purpose of positioning a moving object in the building, the error is tolerable.

5 Conclusions

An Emergency Deployment System based on multiple-communications. For practical alarm system model is designed and some experiments for checking the performance of unit technologies are performed.

For the multiple communication methods, energy consumptions of 1 to N one-way communication for sending one way codes and communication method for guiding and tracking the responders have been analyzed. The GCM Push and BLE communications are preferred for sending an emergency one-way notification to conserve power. Wi-Fi communications are preferred for two-way communications among the server and clients because a large amount of data exchange is required in this process.

The effects of indoor positioning based on fingerprint is analyzed. The experiment result shows that the proposed system can be used as practical one.

For improving accuracy of positioning, some better methods have to be studied. One candidate method is UWB (Ultra Wide Band) for better accurate positioning decision.

Acknowledgement. This research was supported by The Leading Human Resource Training Program of Regional Neo industry through the National Research Foundation of Korea (NRF-2016H1D5A1909408) funded by the Ministry of Science, ICT and future Planning, and the BK21 Plus Program funded by the Ministry of Education (MOE, Korea) and National Research Foundation of Korea (NRF).

References

1. Ahn, H., Lee, P., Hong, J.: An algorithm of localization by using the score of multiple power beacon signals in wireless sensor networks. Korea Comput. Congr. **37**(1)(D), 165–169 (2010)
2. Stallings, W.: Data & Computer Communication, 10th edn. Pearson Education, Inc. (2014)
3. Kong, In-Yeup, Kim, Ho-Jin: Experiments and its analysis on the identification of indoor location by visible light communication using LED lights. J. Korea Inst. Inf. Commun. Eng. **15**(5), 1045–1052 (2011)
4. Vasisht, D., Kumar, S., Katabi, D.: Decimeter-level localization with a SingleWiFi access point. In: 13th USENIX Symposium on Networked Systems Design and Implementation (NSDI '16)
5. Pathak, O., Palaskar, P., Palkar, R., Tawari, M.: Wi-Fi indoor positioning system based on RSSI measurements from Wi-Fi access points – a tri-lateration approach. Int. J. Sci. Eng. Res. **5**(4), April 2014
6. WL18xxMOD WiLink™ 8 Single-Band Combo Module – Wi-Fi®, Bluetooth®, and Bluetooth Low Energy (LE). http://www.ti.com
7. Google Cloud Messaging (GCM) Push. https://developers.google.com/cloud-messaging/
8. Wireless LAN-Based LBS Services, Hakyong KIM, April 2006
9. Bluetooth Indoor Localization System, Gunter FISCHER, Burkhart DIETRICH, Frank WINKLER
10. Dong, Q., Dargie, W.: Evaluation of the Reliability of RSSI for Indoor Localization, August 2012
11. Park, J., Jung, M.A., Yoon, S., Lee, S.R.: System design for location determination inside the ship. J. Korean Inst. Commun. Sci. **38**(2), 181–188 (8 p.) (2013)
12. Hatami, A., Pahlavan, K.: Comparative statistical analysis of indoor Positioning using empirical data and indoor radio channel models. In: IEEE Communications Society Subject Matter Experts for Publication in the IEEE CCNC 2006 Proceedings, pp. 1018–1022 (2006)
13. Heo, S.J., Cho, C.H., Kim, J.Y.: Advertising implications of beacon technology - focusing on tam to predict user acceptance of beacon application. Korean J. Advert. Public Relat. **17**(3), 98–137 (2015)
14. Smith, T.F., Waterman, M.S.: Identification of common molecular subsequences. J. Mol. Biol. **147**, 195–197 (1981)

Network Engineering: Towards Data-Driven Framework for Network Configuration

Khac Hoai Nam Bui, Sungrae Cho, Jason J. Jung$^{(\boxtimes)}$, Joongheon Kim,
O-Joun Lee, and Woongsoo Na

Department of Computer Engineering, Chung-Ang University, Seoul 156-756, Korea
hoainam.bk2012@gmail.com, {srcho,joongheon}@cau.ac.kr, j2jung@gmail.com,
concerto9203@gmail.com, wsna@uclab.re.kr

Abstract. In this paper, we want to introduce a new research area "network engineering". The main research question is how the network configuration can be automatically and adaptively decided, given various dynamic contexts (e.g., network interference, heterogeneity and so on). The aim of this work is to design data-driven framework which is in three layer architecture (i.e., network entity layer, complex semantic analytics layer, and action provisioning layer).

Keywords: Network ontology · Network engineering · Data science
Automated-Provisioning

1 Network Engineering as a New Research Area

With the developing of Internet of Things (IoT), there have been a large number of networking technologies which enable wired/wireless electronic devices to communicate with each other (Fig. 1). However, the most important development trends of IoT is integrating with existing network system. For instance, given various dynamic contexts (e.g., network interference, heterogeneity and so on), it has been difficult to decide how to build the appropriate configurations for heterogeneous network.

As technology progresses, more and more processing power, storage and battery capacity become available at relatively low cost and with limited space requirements [1]. This trend is enabling the development of extremely small-scale electronic devices with identification, communication, and computing capabilities, which could be embedded in the environment or in common objects. The development of such a new class of services will, in turn, require the introduction of novel paradigms and solutions for communications, networking, computing and software engineering.

In this paper, we introduce a new research area "network engineering". Summarizing, the aim of our research focuses on key system-level issues which are defined as follows:

© ICST Institute for Computer Sciences, Social Informatics and Telecommunications Engineering 2018
J. J. Jung et al. (Eds.): BDTA 2017, LNICST 248, pp. 60–65, 2018.
https://doi.org/10.1007/978-3-319-98752-1_7

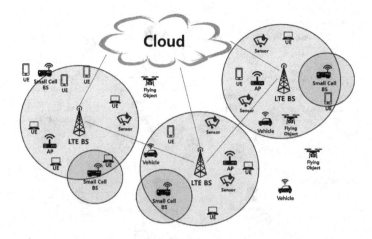

Fig. 1. IoT-based heterogeneous network

- *Devices Heterogeneity*: IoT will be characterized by a large heterogeneity in terms of devices taking part in the system, which are expected to present very different capabilities from the computational and communication standpoints. The management of such a high level of heterogeneity shall be supported at both architectural and protocol levels [2].
- *Network Interference and Scalability*: When large numbers of devices are deployed in urban environments where the ISM bands are already overcrowded the interference from external sources. But how can we allow the devices to really talk to each other without increased signaling in the network and long delays? How can we be sure that when an alarm is raised by an IoT device, this information will be prioritized and sent immediately to the respective target device without being lost due to collisions or interference? they are research questions that we take into account in this work [3,4].
- *Service Provisioning and Management*: due to the massive number of services/service execution options that could be available and the need to handle heterogeneous resources.

2 Data-Driven Framework for Network Configuration

The typical IoT network comprises thousands of connected device using different protocols which have various resources, complex interdependencies and security requirements. Traditional analysis techniques are not able to deal with the configuration challenges of heterogeneous network in terms of scalability, interoperability and security. In this regard, a novel data-driven framework is required to semantically model the network configuration and automated provisioning.

Figure 2 shows an example for inferring the response method for an upset condition of Network based on our proposed framework. In particular, our research context on the development of Data-Driven Framework for network configuration combining three layer architecture as follows:

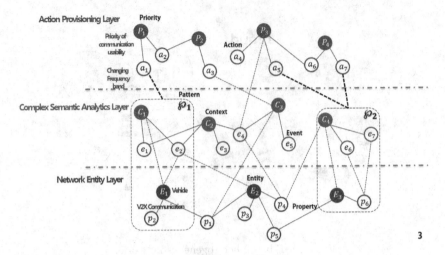

Fig. 2. An example for inferring the response method/action for an upset condition of network

2.1 Network Entity Layer

In this layer, we take into account defining the information of network entities (e.g., Mobile Devices, Sensors, Connected Vehicles and so on) in terms of situation, prior knowledge and association rule. In particular, few features should be properly accounted for:

- Establishing a classification system for entities on the network
- Establishing multi-layer network ontology model
- Describing the network entities, the situations and existing network configuration methodologies
- Inter-layer relationship technology on multi-layer network ontology
- Existing network ontology survey and ontology integration methodology
- Gradual expansion of knowledge base through network ontology integration

2.2 Complex Semantic Analytics Layer

This layer focuses on Network Learning and Context Awareness Methodologies in term of developing functions that automatically detects the situation and environment including network condition, environmental factor, user's request, service operator's requirement. The components and dimensions include:

- Establishing network phenomenon event sensing model
- Detecting methods of Abnormality
- Establishing window size of determination method according to status abnormality
- Abnormal pattern modeling and pattern library construction
- Status fault pattern establishment of high-speed detection method
- Developing event detection model and pattern library learning method

2.3   Action Provisioning Layer

Regarding the application layer, we focus on developing Adaptive Network Provisioning Methodology Module that can be applied to various network domains such as automatic, resource allocation, retrieval, relocation for optimized system resources based on perceived state and system configuration. The research context are defined as follows:

- Development of abnormal pattern analysis methodology
- Network state inference engine development
- Development of evaluation method for Action candidates
- Development of evaluation method for performance result of action
- Development of inference engine learning method based on performance result

3   Adaptive and Automated Provisioning System

Current provisioning and configuring networks are manually intensive processes focused on individual, vendor-specific, network elements rather than the holistic provisioning of data centers across distributed networks and virtual environments. These manual configurations are not able to keep up with rapidly changing devices and networks, creating outage risks for network and data center that forfeit revenue, customer trust, and delay the introduction of new services.

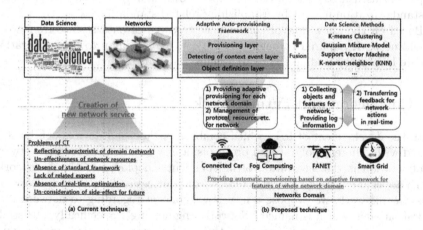

Fig. 3. Adaptive and Automated Provisioning system based on data analysis

In this regard, the objective of this research is developing an Adaptive Automated-Provisioning (AAP) framework for network engineering that can be applied to a variety of network platforms in preparation for the hyper-connected network revolution era. Figure 3 shows the proposed techniques that we consider to develop the AAP framework to improve the capacity of the network based on big data analysis.

4   Discussion and Related Work

As the number of connected things is rapidly growing, the network configuration need to transform in terms of concepts, architecture and protocols. Hence, a dynamic global network infrastructure with self-configuring capabilities becomes a hot issue in this research area. In this study, we introduce a new novel concept, "Network Engineering" which focuses on big data analysis for network configuration in case of automated and adaptive system. Our expected contributions of this study as follows:

- It relieves the user's trust and contributes to the improvement of the national status by eliminating the side effects of the wireless network market applying various algorithms based on data science or artificial intelligence [5–7].
- A system that can cope with various network phenomenon events such as shortage of wireless network resources, link disconnection due to channel and network node mobility, and so on in real time, contributes to enhance the quality of life of the people [8,9].
- Various network services can be created through various wireless network domains (e.g., Smart Grid Platform, Connected Car, Fog Computing, 5G Networks) and more intelligent and stable service can be provided [10–13].
- It is anticipated that it will provide a foundation to lead the global market by supplying low-priced standardized frameworks to domestic and overseas companies/institutes. In addition, we expect to be able to lead the international standardization market by establishing related standardization TG.
- Based on the know-how of convergence research of data science framework and network acquired through this project, we can jump to next-generation core laboratory leading the related fields and emit related research experts.

5   Conclusion and Future Work

With the development of IoTs which combines a thousand devices can be connected with different policies and protocols, a new requirement for the Network Configuration is the integration of independently deployed IoTs sub-networks which are characterized by very heterogeneous devices and connectivity capabilities. In this work, we propose a new novel concept regarding to network configuration area, which is named "Network Engineering". Specifically, our study focuses on improving and increasing the network capacity, performance gain, automated and adaptive provisioning in heterogeneous network environments based on data analysis techniques.

For the next steps of this work, we take into account developing and providing an Adaptive and Automated Provisioning (AAP) framework for Heterogeneous Network which include centralized, decentralized and hybrid systems. Evaluating the advantages of each system with the aim is optimizing capacity of the total network.

Acknowledgments. This work was supported by the National Research Foundation of Korea (NRF) grant funded by the Korea government (MSIP) (NRF-2017R1A41015675).

References

1. Gupta, A., Jha, R.-K.: A survey of 5G network: architecture and emerging technologies. IEEE Access **3**, 1206–1232 (2015)
2. Miornandi, D., Sicari, S., Pellegrini, F.-D., Chlamtac, I.: Internet of Things: vision, applications and research challenges. Ad Hoc Netw. **10**(7), 1497–1516 (2012)
3. China Mobile C-RAN: The road towards green RAN. White Paper, Ver, 2 (2011)
4. Tragos, E.-Z., Angelakis, V.: Cognitive radio inspired M2M communications. In: Proceedings of 16th IEEE International Symposium on Wireless Personal Multimedia Communications (WPMC), pp. 1–5 (2013)
5. Gubbi, J., Buyya, R., Marusic, S., Palaniswami, M.: Internet of Things (IoT): a vision, architectural elements, and future directions. Futur. Gener. Comput. Syst. **29**(7), 1645–1660 (2013)
6. Bhargav, A.-S., Squicciarini, A.-C., Bertino, E.: Trust negotiation in identity management. IEEE Secur. Priv. **5**(2), 55–63 (2007)
7. Nguyen, H.-L., Lee, O.-J., Jung, J.-E., Park, J., Um, T.-W., Lee, H.-W.: Event-driven trust refreshment on ambient services. IEEE Access **5**, 4664–4670 (2017)
8. Zorzi, M., Gluhak, A., Lange, S., Bassi, A.: From today's Intranet of Things to a future Internet of Things: a wireless-and mobility-related view. IEEE Wirel. Commun. **17**(6), 43–51 (2010)
9. Yu, G.-J., Bui, K.-H.-N.: A novel downlink interference management mechanism for two-tier OFDMA femtocell networks. J. Adv. Comput. Netw. **4**(2), 80–85 (2016)
10. Atzori, L., Iera, A., Morabito, G.: The The Internet of Things: a survey. Comput. Netw. **54**(15), 2787–2805 (2010)
11. Yun, M., Yuxin, B.: Research on the architecture and key technology of Internet of Things (IoT) applied on smart grid. In: Proceedings of IEEE International Conference on Advances in Energy Engineering (ICAEE), pp. 69–72 (2010)
12. Bui, K.-H.-N., Camacho, D., Jung, J.-E.: Real-time traffic flow management based on inter-object communication: a case study at intersection. Mob. Netw. Appl. **22**(4), 613–624 (2017)
13. Bui, K.-H.-N., Jung, J.-E., Camacho, D.: The game theoretic approach on real-time decision making for IoT-based traffic light control. Concurr. Comput. Pract. Exp. **29**(1511), e4077 (2017)

Detecting Human Emotion via Speech Recognition by Using Ensemble Classification Model

Sathit Prasomphan[✉] and Surinee Doungwichain

Department of Computer and Information Science, Faculty of Applied Science,
King Mongkut's University of Technology North Bangkok,
1518 Pracharat 1 Road, Wongsawang, Bangsue, Bangkok 10800, Thailand
ssp.kmutnb@gmail.com

Abstract. Speech Emotion Recognition is one of the most challenging researches in the field of Human-Computer Interaction (HCI). The accuracy of detecting emotion depends on several factors for example, type of emotion and number of emotion which is classified, quality of speech. In this research, we introduced the process of detecting 4 different emotion types (anger, happy, natural, and sad) from Thai speech which was recorded from Thai drama show which was most similar with daily life speech. The proposed algorithms used the combination of Support Vector Machine, Neural Network and k-Nearest Neighbors for emotion classification by using the ensemble classification method with majority weight voting. The experimental results show that emotion classification by using the ensemble classification method by using the majority weight voting can efficiency give the better accuracy results than the single model. The proposed method has better results when using with fundamental frequency (F0) and Mel-frequency cepstral coefficients (MFCC) of speech which give the accuracy results at 70.69%.

Keywords: Speech emotion recognition · Feature extraction
Ensemble classification · Weight majority vote · k-nearest neighbor
Neural Network · Support Vector Machines

1 Introduction

Speech Emotion Recognition (SER) is a challenging research area in the field of Human-Computer Interaction (HCI). The purpose of SER is to recognize emotion such as anger, disgust, fear, happiness, sadness, etc. from tonal variations in human speech [1, 2]. Several algorithms were introduced to make computer to be able to understand and to be able to classify several types of emotion in human speech. Some benefit of knowing this emotion from speech is to use with the application which requires a man-machine interaction such as computer tutorial, automatic translation, mobile interaction, health care, children education, etc. Emotion is an importance mental and physiological state. In natural, baby learns to recognize emotional information before understanding semantic information in his/her mother's utterance [3]. Reliable emotion detection in usability tests will help to prevent negative emotion [4]. Detecting emotion

© ICST Institute for Computer Sciences, Social Informatics and Telecommunications Engineering 2018
J. J. Jung et al. (Eds.): BDTA 2017, LNICST 248, pp. 66–73, 2018.
https://doi.org/10.1007/978-3-319-98752-1_8

can help particularly for user opinion mining or stress prevention [3, 4]. Computer may not be able to exactly understand the natural of these emotions unless we employ the speech processing. Many researchers have used the statistics of difference speech attributes for being a representation of each sound such as pitch, formant, amplitude or power of the speech. Speech features can be classified to one of these three categories: *prosodic features* such as pitch (F0), intensity and duration, *voice quality* and *spectral features* such as Mel-Frequency Cepstral Coefficients (MFCC) or Linear Prediction Cepstral Coefficients (LPCC).

In case of classification model, researchers offer several model such as Support Vector Machines (SVM) [3, 8], Gaussian Mixture Model (GMM) [2], Hidden Markov Modeling (HMM) [7], k-Nearest Neighbor (k-NN) [10], and Neural Network (NN) [12, 15]. Although the above techniques provided the better classification accuracy, however these techniques are single model that resulted in a used data set in the study must include the parameter configuration step. Moreover, each parameter must be fixed which cause to bias and poor performance. Another way to reduce bias is to use common decision (ensemble), which can create diversity and minimize the errors caused by the variance [9]. Researchers have attempted to make ensemble decision applied to enhance the emotion classification. Anagnostopoulos et al. [11] presented a research on *ensemble majority voting classifier* for speech emotion by using a decision from the base classifier. The decision with majority voting using k-NN, C4.5, and SVM with polynomial kernel was used to find a suitable model to classify the speech in HUMAINE database [5]. These framework provided accuracy by 96%. Morrison [14] presented a technique for searching feature that combined ensemble model by using the base classifier SVM with RBF kernel, random forest, k-NN, K* and multilayer perceptron. The algorithm provided accuracy by 79.43% and 73.29% for NATURAL and ESMBS database in ordering. The results showed that if a dataset has different types of information and emotion, the feature selection methods and ensemble model will be different. Vasuki [17] focused on searching frame work to reject noisy and weak input file by using the weight factor ensemble model with SVM classifier to detect outliers. If input is unusual, it will be rejected from the training dataset. This framework showed the accuracy by 74.70%.

From all of these research shows that the ensemble model can increase the performance of emotion classification from speech. It also emphasizes that the effectiveness of methods for separating emotion depends on several factors such as the properties of the selected feature in the experiments, number of emotions; the quality of the audio data is also affected as well. Therefore, we have selected a set of features that are critical to the dataset and methodology to optimize performance of classification by using ensemble model.

The paper is organized as follows. Following this, Sect. 2 provides the details in speech emotion recognition. Section 3 discusses the experimental setup. Results and performance comparison are given in Sect. 4. Section 5 gives conclusions and discussions.

2 Speech Emotion Recognition System

The speech emotion recognition system is described in Fig. 1. In this section, the pre-processing of speech signal which is the pre-emphasis, frame blocking and Hamming windowing was described. The following speech features: energy, zero-crossing rate (ZCR), pitch, MFCC was used. The feature normalization was calculated for every windows of a specified number of frames by statistical method. Classifier was modeled to classify emotions. Finally, the ensemble model was applied to integrate the result of classifier with weighted majority vote. Details of each process can be described as followed:

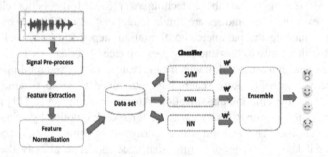

Fig. 1. Speech emotion recognition system.

A. Signal Pre-processing: The basic operations used in the speech pre-processing include the following: pre-emphasis, frame blocking and hamming windowing.

1. Pre-emphasis: The speech signal $s(n)$ is sent to a high-pass filter as show in Eq. (1).

$$s2(n) = s(n) - a * s(n - 1) \qquad (1)$$

where $s2(n)$ is the output signal and the value of a is usually between 0.9 and 1.0. The z-transform of the filter is given in Eq. (2).

$$H(z) = 1 - a * z - 1 \qquad (2)$$

The goal of pre-emphasis is to compensate the high-frequency part that was suppressed during the sound production mechanism of humans. Moreover, it can amplify the importance of high-frequency formants.

2. Frame blocking: The speech signal is divided into a sequence of frames where each frame can be analyzed independently and represented by a single feature vector. Frame shift is the time difference between the start points of successive frames, and the frame length is the time duration of each frame. The frame block is of length 10 ms to 40 ms from the filtered signal at every interval of 1/2 or 1/3 of frame length.

3. Hamming windowing: In order to keep the continuity of the first and the last points in the frame. If the signal in a frame is denoted by $s(n), n = 0, \ldots N - 1$

Then the signal after Hamming windowing is $s(n) * w(n)$, where the Hamming window $w(n)$ is defined in Eq. (3).

$$w(n, a) = (1 - a) - a \cos(2pn/(N - 1)) \tag{3}$$

B. Feature Extraction: The speech feature extraction which is also called speech coding is a very important and it is basically part in the automatic speech processing systems. Features of the speech are generally obtained from the digital speech. Various methods are utilized that aim to extract speech features which are useful to classify the type of emotions. In this research, the extracted features are energy, ZCR, pitch, and MFCC.

C. Feature Normalization: The speech segments have different lengths. In order to obtain isometric speech segments and reduce redundancy of data, the statistical method [3] was adopted to normalize the states. For each coefficient, mean, variances, median, maximum and minimum across all frames are, calculated.

D. Classifiers: Classifier is another component of a speech emotion recognition system. In this research, we used three classification methods: SVM, Neural Network and k-nearest neighbor.

E. Ensemble Classification Method: Ensemble classifier is the model which combines several classifiers' technique for solving the same problem by using the results from all of classifiers for decision in the final step. Ensemble model [13, 16] composed of several model for example, vote ensemble which uses the same training data with several classifier, bootstrap aggregating (Bagging) which uses the random training data and constructs the single ensemble model, and random forest which similar to bagging technique but instead of using random data, it randomly selects attribute from dataset and uses several decision tree for becoming classifier in the ensemble model. In this research, the vote ensemble with base classifiers which has low computational complexity and difference theoretical background was selected. The proposed model aims to reduce bias and redundancy [11] by using the combined model with weighted majority vote. If the classifier in the ensemble does not provide the identical classification result, then it is reasonable to attempt to give the more competent classifiers more power in making the final decision. We called this step is weighted majority vote. The formula for weighted majority vote is shown in Eq. (4).

$$\sum_{t=1:T} w_t \; d_t, \; J(x) = max_{j=1,...,c} \, m \sum_{t=1:T} w_t \; d_{t,j} \tag{4}$$

The T classifiers are class-conditionally independent with accuracies $p_1, \ldots, p_T$. The optimal weights for the weighted majority voting rule can be shown to be

$$w_t = p_t/(1 - p_t) \tag{5}$$

3   Experimental Setup

Thai Emotional speech corpus [6] was used to classify emotion states. This corpus construction has been funded by National Electronics and Computer Technology Center (NECTEC). All emotional speech collected from conversations by professional actors and actresses in a Thai drama show that contains many background music and noise within the speech. There are two groups of emotion were used to annotate in this corpus. The first group consists of four basic emotions: neutral, happy, sad, and angry. The second group consists of twelve labels: happiness, satisfaction, fear, surprise, anger, jealousy, rage, doubt, hate, excitement, sadness, and fun. For this research, we firstly focused on detecting emotions from the first group. It is possible to recognize four real emotions of human. We used only 352 utterances from 2908 utterances in the corpus were utilized in this work. The details of each emotion are shown in Table 1.

Table 1. Number of emotions in Thai emotion speech corpus.

Emotion	Male	Female	Male + Female
Anger	47	72	119
Happy	42	40	82
Neutral	41	40	81
Sad	33	37	70
Total	163	189	352

In Thai emotion speech corpus, we have randomly selected 352 speeches for study which with and without noise, background music or one of them for the diversity of speech in the experiment. We use the cross-validation with holdout 1/3 to split the data into two sets for training and testing, 236 training speech and 116 testing speech.

In case of signal pre-process, the coefficient was set in the pre-emphasis step with 0.9375. In the framing process, frame has been segmented with size of 480 samples or approximately 30 ms, and the distance between the frames (frame overlap) is 240 samples or about 15 ms. After that we used Hamming window to emphasize the importance signal in the middle frame signal. The speech feature used in this research was energy, ZCR, F0 and MFCC. After feature extraction had been processed, the feature was normalized by using statistical methods. The important features were combined to analyze if it most affects to emotional classification. We used MFCC to combine with *prosodic feature* (energy, ZCR, F0) due to the MFCC feature give the highest accuracy compared to prosodic feature as shown in Table 2.

Emotion classification has been created by using the ensemble model from the same set of data. When each classifier gives the predicted class, these results will be weight for each classifier which is [2, 6, 7] for SVM with RBF-7 kernel function, KNN, and NN. The setting weight values depend on the prediction accuracy. After that, the sum of predicted class and predicted weight in each classifier was calculated for voting. The performance of proposed model was based on evaluation of data classification performance by using Eq. (6),

Table 2. The Classification accuracy in different features and models.

No.	Feature	No. feature	Accuracy (%)				
			SVM (RBF 7)	KNN	NN	Bagging	Weighted majority vote
1	F0	5	40.52	43.10	40.52	40.52	42.24
2	Energy	5	35.34	42.24	50.00	42.24	40.52
3	ZCR	5	41.38	37.93	40.52	41.38	41.38
4	MFC	105	62.93	56.03	58.62	60.34	68.97
5	MFCC + F0	110	**66.38**	56.90	61.21	62.93	**70.69**
6	MFCC + Energy	110	65.52	56.03	62.07	61.21	66.38
7	MFCC + ZCR	110	62.07	56.03	**65.52**	**63.79**	66.38
8	MFCC + F0 + Energy	115	**66.38**	57.76	62.07	60.34	68.10
9	MFCC + F0 + ZCR	115	65.52	52.59	58.62	58.62	69.83
10	MFCC + Energy + ZCR	115	61.21	**62.07**	59.48	**64.66**	65.52
11	MFCC + F0 + Energy + ZCR	120	63.79	59.48	57.76	62.07	66.38

$$Accuracy = (TP + TN)/(TP + FN + TN + TN) * 100 \qquad (6)$$

where, TP is true positive, TN is true negative, FP is false positive, FN is false negative.

4   Experimental Result

In this research, 5 models were tested for the speech classification accuracy which is 3 single model: SVM with RBF-7 kernel function, k-NN, and NN, and 2 ensemble models: bagging which uses base classifier by using decision tree and weighted majority vote with 3 base classifier model: SVM with RBF-7 kernel function, k-NN and NN. In addition, the feature was also compared its classification accuracy.

From Table 2, it shows that model which can give the best classification accuracy for speech emotion classification is ensemble weighted majority vote by using F0 and MFCC.

The confusion matrix in Table 3 showed that ensemble weighted majority vote model with F0 and MFCC give the best accuracy with 70.69%.

Table 3. Confusion matrix for the feature set MFCC + F0 of ensemble weighted majority vote.

Emotion	Recognized emotions (%)			
	Anger	Happy	Neutral	Sad
Anger	**76.92**	7.69	12.82	2.56
Happy	14.81	**66.67**	14.81	3.70
Neutral	11.11	18.52	**62.96**	7.41
Sad	8.70	0.00	17.39	**73.91**

5 Conclusions

This research presents a novel algorithm for detecting human emotion via speech recognition by using ensemble classification model. The proposed algorithm aims to detect the emotional by using information with the combination of SVM classifier, Neural Network classifier and k-Nearest Neighbor with the weighted majority voting ensemble method with combine speech feature Fundamental Frequency (F0) and Mel Frequency Cepstral Coefficient (MFCC) for Thai emotional speech corpus. The experimental results show that the proposed framework can efficiently find the correct speech emotion compared to by using the comparing method. For the future work, the process for noise removal and background music should be considered. In addition, the feature selection and model selection for improve the classification accuracy should be focused.

Acknowledgment. This research was funded by King Mongkut's University of Technology North Bangkok. Contract no. KMUTNB-58-GEN-048.

References

1. Ayadi, M.M.H.E., Kamel, M.S., Karray, F.: Survey on speech emotion recognition: features, classification schemes, and databases. Pattern Recognit. **44**, 572–587 (2011)
2. Xu, S., Liu, Y., Liu, X.: Speaker recognition and speech emotion recognition based on GMM. In: 3rd International Conference on Electric and Electronics (2013)
3. Seehapoch, T., Wongthanavasu, S.: Speech emotion recognition using Support Vector Machines. In: the 5th International Conference on Knowledge and Smart Technology (KST), pp. 219–223 (2011)
4. Stickel, C., Ebner, M., Steinbach-Nordmann, S., Searle, G., Holzinger, A.: Emotion detection: application of the valence arousal space for rapid biological usability testing to enhance universal access. In: Stephanidis, C. (ed.) UAHCI 2009. LNCS, vol. 5614, pp. 615–624. Springer, Heidelberg (2009). https://doi.org/10.1007/978-3-642-02707-9_70
5. Burkhardt, F., Paeschke, A., Rolfes, M., Sendlmeier, W., Weiss, B.: A database of German emotional speech. In: Proceedings of Interspeech (2005)
6. Kasuriya, S., Teeramunkong, T., Wutiwiwatchai, C.: Developing a Thai emotional speech corpus. In: International Conference on Asian Spoken Language Research and Evaluation (2013)
7. Kasuriya, S., Banchaditt, T., Somboon, N., Teeramunkong, T., Wutiwiwatchai, C.: Detecting emotional speech in Thai drama. In: 2nd ICT International Student Project Conference (ICT-ISPC) (2013)
8. Shen, P., Changjun, Z.: Automatic speech emotion recognition using support vector machine. In: International Conference on Electronic & Mechanical Engineering and Information Technology, pp. 621–625 (2011)
9. Thamsiri, D., Meesad, P.: Ensemble data classification based on decision tree, artificial neuron network and support vector machine optimized by genetic algorithm. J. King's Mongkut's Univ. Technol. North Bangk. **21**(2), 293–303 (2011)
10. Rieger Jr., S.A., Muraleedharan, R., Ramachandran, R.P.: Speech based emotion recognition using spectral feature extraction and an ensemble of kNN classifiers. In: 9th International Symposium on Chinese Spoken Language Processing (ISCSLP), pp. 589–593 (2014)

11. Anagnostopoulos, T., Skourlas, C.: Ensemble majority voting classifier for speech emotion recognition and prediction. J. Syst. Inf. Technol. **16**(3), 222–232 (2014)
12. Nicholson, J., Takahashi, K., Nakatsu, R.: Emotion recognition in speech using neural networks. In: 6th International Conference on Neural Information Processing, vol. 2, pp. 495–501 (1999)
13. Mu, X., Lu, J., Watta, P., Hassoun, M.H.: Weighted voting-based ensemble classifiers with application to human face recognition and voice recognition. In: Proceedings of International Joint Conference on Neural Networks, Atlanta, Georgia, USA, 14–19 June, pp. 2168–2171 (2009)
14. Morrison, D., Wang, R., De Silva, L.C.: Ensemble methods for spoken emotion recognition in call-centres. J. Speech Commun. **49**(2), 98–112 (2007)
15. Aha, D., Kibler, D.: Instance-based learning algorithms. Mach. Learn. **6**, 37–66 (1991)
16. Sharkey, A.J.C.: Combining Artificial Neural Nets. Ensemble and Modular Multi-Net Systems. Springer, London (1999). https://doi.org/10.1007/978-1-4471-0793-4
17. Vasuki, P.: Speech emotion recognition using adaptive ensemble of class specific classifiers. Res. J. Appl. Sci. Eng. Technol. **9**(12), 1105–1114 (2015)

Pattern Analysis of Natural Disasters in the Philippines

Marie Elaine Joyce N. Garcia(✉) and Alexander A. Hernandez

College of Information Technology Education,
Technological Institute of the Philippines, Manila, Philippines
marieelainejoycengarcia@gmail.com,
alexander.hernandez@tip.edu.ph

Abstract. The Philippines is one of the most vulnerable countries to natural disasters due to its geographical location. Natural disasters such as storms, floods, earthquakes, and droughts often occur which bring threat and disturbance to people. Its frequency and severity are probable to increase since there are many studies linking its relationship with climate change. This study analyzed the frequency of natural disasters occurred from the dataset provided by the CRED EM-DAT to identify the patterns from a specified period from 1980–2012. The Results revealed that there is a drastic increase in the frequency of storms that hit the Philippines for more than three (3) decades from 1980 to 2012. Storms, floods, and volcanic eruptions remain the top 3 natural disasters that affect the entire country. There is an increase of 147% from 1980–2012. Thus, the natural disasters are extensively aggravating several industries in the Philippines. The results also indicate that the frequency of natural disasters is more likely increase in the future.

Keywords: Data mining · Natural disasters · Clustering · Developing country

1 Introduction

The increasing number of occurrences of natural disasters has always been one of the main concerns of governments worldwide since its frequency and intensity are expected to rise even further shortly [1]. A natural disaster is defined as the actual happening of a natural event that causes significant disturbance and loss [2] such as alarm to the public [3], harm to people and damages to properties [4] It is caused by geophysical, meteorological, hydrological, climatological, extra-terrestrial and biological phenomena, which adversely impact the affected areas' environment [5]. Examples of natural disasters are storms, earthquakes, droughts, hurricanes, heat waves, thunderstorms and lightning [6].

Emergency Events Database or EM-DAT, which is governed by the Centre for Research on the Epidemiology of Disasters (CRED), provides that for hazard to qualify as a natural disaster it must fulfill at least one of the following criteria: (a) ten (10) or more people reported killed; (b) hundred (100) or more people reported affected; (c) declaration of a state of emergency and a; (d) call for international assistance [6].

© ICST Institute for Computer Sciences, Social Informatics and Telecommunications Engineering 2018
J. J. Jung et al. (Eds.): BDTA 2017, LNICST 248, pp. 74–83, 2018.
https://doi.org/10.1007/978-3-319-98752-1_9

Natural disasters are also associated with climate change. According to many studies, climate change has a strong relationship to the occurrences of natural disasters since it increases the frequency and severity of the latter [7–9]. The National Aeronautics and Space Administration (NASA) lay down all the probable consequences of the climate change which include, among others the increase in droughts and heat waves; changes in precipitation patterns and; stronger and more intense hurricanes. [10]

For the past few decades, many countries have experienced immense and distressing natural disasters [11]. The United Nations Office for Disaster Risk Reduction (UNISDR) in its report entitled, "The Human Cost of Weather-Related Disasters", provided that ninety percent (90%) of the disasters have been caused by floods, storms, heat waves and other weather-related events which claimed 606,000 lives, with additional 4.1 billion people injured, left homeless or in need of emergency assistance [12]. Some of the recorded major natural disasters in the planet are (1) Hurricane Katrina in 2005 which quickly moved to the city of New Orleans and resulted in massive flooding, loss of life and property. [13]; (2) Indian Ocean tsunami in 2004 which have received worldwide media coverage [14]; (3) the heavy monsoon that hit Pakistan in 2010 which also caused floods that affected the country, bringing immense damage to homes, schools, fields, and infrastructure. [7]; (4) the gigantic earthquake in Tohoku region, Northeast Japan with magnitude 9.0, followed by a giant tsunami. The tsunami was also historical as its run-up height reached over 39 m. As of early May 2011, over 24 thousand people were reported as dead or missing [15]; and (5) the 2010 earthquakes wrought in Haiti and Chile. The January 2010 earthquake that struck Haiti's densely populated capital, Port-au-Prince, caused significant loss of human life (between 200,000 and 250,000 fatalities), the displacement of hundreds of thousands more, and severe damage to the country's economic infrastructure [16].

In the Philippines, natural disasters such as floods, tropical storms, and droughts also occur frequently [17] due to its geographical setting and physical environment [18, 19]. For the last decade, tropical storms and cyclones often accompanied by storm surges, high winds, flooding, and landslides have caused tremendous disasters [20]. One of the most disastrous natural events ever recorded in the history of the Philippines is the typhoon Haiyan (Yolanda) in 2013 with maximum sustained winds reaching 315 kph with gusts up to 379 kph just before landfall [21]. Over a 16-hour period, the Category 5 super typhoon straightforwardly swept through six provinces in the central Philippines [22]. It was an enormous and extremely intense tropical cyclone that struck the Philippines causing catastrophic damage. According to statistics as of December 12, there were 5,982 fatalities, with an additional 1,799 missing and 27,022 injured [23]. In addition to the loss of life, Typhoon Haiyan also caused extreme economic losses; the damage to infrastructure; communication lines and; agriculture damages were estimated at US$802 million [22, 23].

As historical records show, the damages brought by natural disasters are threats to society claiming and affecting lives and environment. Many studies focus on social and natural sciences, even on economic perspectives with a fairly number of researches [14, 24–27]. There are also studies which map the real-time crisis of natural disasters using social media and public behavior during the occurrence of natural disasters [28, 29]. This research, in counterpart, analyses the patterns of natural disasters occurred in the Philippines using data mining techniques to determine the changes in the frequency that happened over a specified period in the Philippines. It is organized as follows: Part II explains the overview of the Philippine geographical location and its natural disaster profile. It also explains the relationship between climate change and natural disasters; types of natural disasters and other studies that focus on pattern analysis using natural disaster events dataset. Part III presents the techniques used to form the pattern of natural disasters. These include among others, the basic statics, classifying and clustering. Part IV finally explains the conclusion as a result of the pattern analysis.

2 Literature Review

2.1 Overview of Philippine Geographical Location and Its Natural Disaster Profile

The Republic of the Philippines is a sovereign state in archipelagic Southeast Asia, with 7,107 islands spanning more than 300,000 square kilometers of territory. It is divided into three island groups: Luzon, Visayas, and Mindanao [30]. The archipelago has a total land area of 299,000 km 2 which makes it the 64th largest country in the world. It is part of western-Pacific "Ring of Fire" with a total 37 mountains, of which 18 are active. It also has a remarkably long coastline as regards its overall land size of which 67 out of 81 of its provinces have seashore [31].

Due to its exceptionally complex geographical location, the Philippines is vulnerable to natural disasters. Powerful typhoons develop many times per year accompanied by strong winds. The tectonic activity underlying the islands may trigger tsunamis and other deadly threat from the sea [31]. According to Guha-Sapir et al. (2012), the Philippines is one of the top 5 countries most frequently hit by natural disasters together with the United States, India, Indonesia and China [32]. This study is also supported by other researches which state that the Philippines is among the hardest hit by natural disasters, particularly typhoons, floods and droughts in Southeast Asia. [33, 34].

2.2 Climate Change and Natural Disasters

The Intergovernmental Panel on Climate Change (IPCC) which is the leading international body for the assessment of climate change defines climate change as a change in the state of the climate over time, whether due to natural variability or as a result of human activity. Another definition from the United Nations Framework Convention on Climate Change (UNFCCC) states that it refers to a change of climate which is directly or indirectly attributed to human activity that alters the composition of the global atmosphere and that is in addition to natural climate variability observed over comparable time periods [35].

Climate change is considered as one of the many threats the world faces nowadays [36–38]. It is probable to lead to a continued increase in frequency and intensity of certain types of natural hazards in several regions of the world [39] despite efforts made to reduce greenhouse gases [30]. The influence of climate change to occurrences of natural disasters is deemed extremely complicated [40]. However, several pieces of evidence are formed to determine the link between these two topics [41]. It is predicted that such climate change will increase the severity and frequency of climate-related disasters like flash floods, surges, cyclones and severe storms [7–9]. Anderson et al. [41] presented the evidence of the effects of climate change in the European setting, among others are; the more intense precipitation, hurricanes rainfall and increased number of occurrences of significant flooding and drought. On the other hand, Leng et al. [42] studied the climate change impacts on droughts in China. According to the study, droughts in some areas of the country for the next few years suggest that it will become more frequent, prolonged and severe.

In the Philippines, the Department of Interior and Local Government (DILG) analyzed the effects of climate change using the available data from 1951 to 2009. Results showed that an average of 20 tropical cyclones formed and cross the Philippines Area of Responsibility (PAR). There was also an increase in some hot days but the decrease of cold nights and the annual mean temperature increased by 0.57 °C [43]. Another study by Garcia et al. [44] states that climate change alters the geographic distribution of forest ecosystems. The study evaluated the effects of climate change on fourteen (14) threatened forest tree species in the Philippines. Out of the total number, 7 or 50% are threatened to decline in their suitable habitat.

Climate change has always been an important concern that the world needs to address since it has a definite relationship to the occurrences of natural disasters. Cases presented by Europe, China and the Philippines only point to one conclusion that is, the effects of climate change is becoming worse in certain parts of the world.

2.3 Pattern Analysis of Natural Disasters

At present, there are three worldwide and multi-peril loss databases such as the NatCatSERVICE (Munich Re), Sigma (Swiss Re) and EM-DAT (Centre for Research on the Epidemiology of Disasters). They focus on national or regional issues on specific hazards and specific sectors. The EM-DAT, in particular, provides for the criteria to be considered as a natural disaster as mentioned earlier [45].

Analysis of natural disasters patterns become very useful since it gives researchers better understanding of the big picture of what is happening. Several studies have analyzed the patterns of natural disasters to validate specific objectives mainly to identify the relationship with climate change [46], its effects on economy [47, 48] mitigation of risks associated with it [49] or merely to determine the annual disaster frequency together with its reported total deaths, affected people, and damages [50]. Most of these studies have determined that based on patterns from historical records, natural disasters can bring negative consequences in the future.

3 Methodology

The methodology used in the study is the Knowledge Discovery (KDD) in discovering the patterns in the dataset [51]. This method allowed the discovery of patterns and evaluated to draw appropriate conclusions. The following describes the dataset and the procedures used in the study: The data from EM-DAT regarding the natural disasters in the Philippines were used in this study. The available data covers the period from 1980 to 2012. This dataset contains, among others, the following: (1) the year the disasters occurred; the location of the disaster; the disaster subgroups and types; latitude and longitude; affected total affected people and; (6) the total damages per U.S. dollars. This study initially cleaned the dataset to eliminate noisy data. Noisy data found in the data set are those records that do not have given values such as dates, places and numbers or amounts. Moreover, sets of values in some columns were also eliminated and not considered to focus on the objective of the research. The attributes that were similar to each other were combined. The attributes selected were carefully assessed to explore on the possible relationships, and present them in an integrated approach. Only relevant data were selected in conjunction with the objective of the research. The cleaned dataset was then classified according to principal categories namely: year, location, disaster subgroup, disaster type and a total number of affected people and total damages per U.S. dollars. The relevant data were selected and transformed into a data set that can be mined. Basic statistics were applied through the use of RapidMiner to gather the number of the most and least observations per category.

4   Results and Discussion

The succeeding figures and tables show the graphical representation produced by RapidMiner and numerical values as provided in items 3.5 to 3.8:

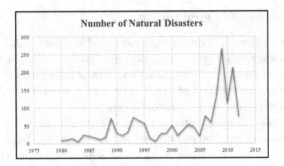

Fig. 1. Annual distribution of natural disasters

As presented in Fig. 1, natural disasters mostly occurred in 2009 followed by 2011. The top five (5) years with most number of natural disasters that occurred in the Philippines were presented in Table 1:

Table 1. Number of natural disasters per year

Year	Number of Natural Disasters
2009	267
2011	214
2008	132
2010	114
2006	78

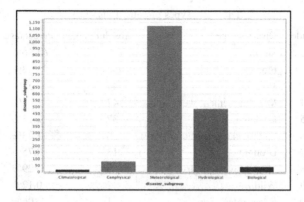

Fig. 2. Disaster subgroup

Figure 2 shows that natural disasters under the Meteorological subgroup represent the highest absolute count of 1,122 or 64.6% of the total recorded occurrences of natural disasters.

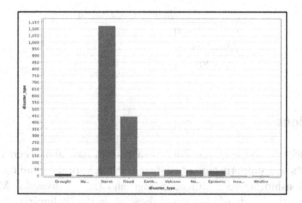

Fig. 3. Disaster types

Figure 3, storms are mostly experienced by the Philippines with an absolute count of 1,122 or 64.56% out of the total given values in the dataset. Further, numerical values of natural disaster types with corresponding percentages were also computed as presented in Table 2:

Table 3 shows that the decade 2001–2012 has a maximum number of days of natural disaster with 119 days. It is due to the bacterial infectious disease happened in Baguio City in 2004. Based on the World Health Organization (WHO), a total of 78 cases of meningococcal disease and 30 deaths were reported from Baguio City and Cordillera Region.

Table 2. Types of natural disasters

Index	Disaster type	Absolute count	Percentage
1	Storm	1,122	64.56%
2	Flood	442	25.43%
3	Volcano	44	2.53%
4	Mass movement wet	42	2.42%
5	Epidemic	37	2.13%
6	Earthquake (seismic activity)	30	1.73%
7	Drought	13	0.75%
8	Mass movement dry	5	0.29%
9	Wildfire	2	0.12%
10	Insect infestation	1	0.06%
Total		**1,738**	**100.00%**

Table 3. Min and max number of days of natural disasters

Decade	Number of days of natural disasters	
	Minimum	Maximum
1980–1990	0	7
1991–2000	0	41
2001–2012	0	119

5 Implications

This study contributes to the current pattern analysis of natural disasters events data. It also supplements and supports current studies in the field of data mining and evidence about the relationship between climate change and natural disasters. Moreover, this study focused on the trends of the natural disasters in the Philippines that may create significant influence in planning disaster risk and reduction management by the national or local government agencies concerned. The patterns produced by Rapid-Miner may be the basis of the national and local agencies and other concerned stakeholders in the formulation of policies and programs to deal with disasters in the Philippines. More importantly, may assist in making safer, adaptive and disaster resilient Filipino communities towards sustainable development.

6 Conclusions and Future Work

Natural disasters are considered one of the problems that the world faces today. Due to its increased frequency and severity, many studies have presented several pieces of evidence to determine the natural disasters' plausible relationship with climate change. As one of the most vulnerable countries in the world, Philippines is frequently hit by natural disasters. Its geographical location is deemed one of the reasons of its explained vulnerability. Data mining techniques were applied to gather pertinent patterns on the dataset provided by the CRED EM-DAT. The results showed that natural disasters drastically increased from 1980 to 2012 by 147% which indicates that the increase of natural disasters experienced by the Philippines is aggravating. The results also suggest that the frequency of natural disasters is likely to increase in the future.

Despite the contributions of this work, the study also has some limitations. First, the study did not explore on the severity of effects of these natural disasters. Thus, it is also necessary to assess the extent of effects on the economy and financial funding provided by the government. Second, this study also did not include developing a predictive model based on the existing weather data attributes. Hence, this study recommends the development of a predictive model to further investigate the long-term effects of natural disasters regarding damages to property, natural resources, and human safety and prevention activities. Finally, this study could be further improved by applying climate analytics models to understand the underlying weather elements and its interactions fully.

Acknowledgments. The authors deeply appreciate the constructive comments and suggestions by the anonymous reviewers of the conference.

References

1. Hoffmann, R., Muttarak, R.: Learn from the past, prepare for the future: impacts of education and experience on disaster preparedness in the Philippines and Thailand. World Dev. **96**, 32–51 (2017)
2. Smith, K.: Environmental Hazards: Assessing Risk and Reducing Disaster. Routledge, Abingdon (2003)
3. Hallegatte, S., Przyluski, V.: The economics of natural disasters: concepts and methods (2010)
4. Noy, I., duPont IV, W.: The long-term consequences of natural disasters—A summary of the literature (2016)
5. Mata-Lima, H., Alvino-Borba, A., Pinheiro, A., Mata-Lima, A., Almeida, J.A.: Impacts of natural disasters on environmental and socio-economic systems: what makes the difference? Ambiente Sociedade **16**(3), 45–64 (2013)
6. Emergency Events Database (EM-DAT)
7. Bergholt, D., Lujala, P.: Climate-related natural disasters, economic growth, and armed civil conflict. J. Peace Res. **49**(1), 147–162 (2012)
8. Visser, H., Petersen, A.C., Ligtvoet, W.: On the relation between weather-related disaster impacts, vulnerability and climate change. Clim. Change **125**(3–4), 461–477 (2014)
9. Miao, Q., Popp, D.: Necessity as the mother of invention: innovative responses to natural disasters. J. Environ. Econ. Manag. **68**(2), 280–295 (2014)
10. National Aeronautics and Space Administration (NASA). The consequences of climate change (2017)
11. Keller, E. A., DeVecchio, D. E.: Natural Hazards: Earth's Processes as Hazards, Disasters, and Catastrophes. Routledge, Abingdon (2016)
12. United Nations Office for Disaster Risk Reduction (UNISDR). The Human Cost of Weather Related Disasters

13. Gunderson, L.: Ecological and human community resilience in response to natural disasters. Ecol. soc. **15**(2) (2010)
14. Cavallo, E., Galiani, S., Noy, I., Pantano, J.: Catastrophic natural disasters and economic growth. Rev. Econ. Stat. **95**(5), 1549–1561 (2013)
15. Mimura, N., Yasuhara, K., Kawagoe, S., Yokoki, H., Kazama, S.: Damage from the great east japan earthquake and tsunami-a quick report. Mitig. Adapt. Strat. Glob. Change **16**(7), 803–818 (2011)
16. Cavallo, E., Noy, I.: Natural disasters and the economy–a survey. Int. Rev. Environ. Res. Econ. **5**(1), 63–102 (2011)
17. Orencio, P.M., Fujii, M.: A localized disaster-resilience index to assess coastal communities based on an analytic hierarchy process (AHP). Int. J. Disaster Risk Reduct. **3**, 62–75 (2013)
18. Pulhin, J.M., Tapia, M.A., Perez, R.T.: Integrating disaster risk reduction and climate change adaptation: Initiatives and challenges in the Philippines. In: Climate Change Adaptation and Disaster Risk Reduction: An Asian Perspective, pp. 217–235. Emerald Group Publishing Limited (2010)
19. Yumul, G.P., Cruz, N.A., Servando, N.T., Dimalanta, C.B.: Extreme weather events and related disasters in the Philippines, 2004–08: a sign of what climate change will mean? Disasters **35**(2), 362–382 (2011)
20. Franta, B., Roa-Quiaoit, H.A., Lo, D., Narisma, G.: Climate Disasters in the Philippines (2016)
21. Lagmay, A.M.F., et al.: Devastating storm surges of typhoon haiyan. Int. J. Disaster Risk Reduct. **11**(1–12) (2015)
22. Lum, T., Margesson, R.: Typhoon Haiyan (Yolanda): US and international response to Philippines disaster. Curr. Politics Econ. South Southeast. Central Asia **23**(2), 209 (2014)
23. National Disaster Risk Reduction and Management Council (2013)
24. Mori, N., Kato, M., Kim, S., Mase, H., Shibutani, Y., Takemi, T., Yasuda, T.: Local amplification of storm surge by Super Typhoon Haiyan in Leyte Gulf. Geophys. Res. Lett. **41**(14), 5106–5113 (2014)
25. Takahashi, B., Tandoc, E.C., Carmichael, C.: Communicating on Twitter during a disaster: an analysis of tweets during Typhoon Haiyan in the Philippines. Comput. Hum. Behav. **50**, 392–398 (2015)
26. Loayza, N.V., Olaberria, E., Rigolini, J., Christiaensen, L.: Natural disasters and growth: going beyond the averages. World Dev. **40**(7), 1317–1336 (2012)
27. Kousky, C.: Informing climate adaptation: a review of the economic costs of natural disasters. Energy Econ. **46**, 576–592 (2014)
28. Middleton, S.E., Middleton, L., Modafferi, S.: Real-time crisis mapping of natural disasters using social media. IEEE Intell. Syst. **29**(2), 9–17 (2014)
29. Chae, J., Thom, D., Jang, Y., Kim, S., Ertl, T., Ebert, D.S.: Public behavior response analysis in disaster events utilizing visual analytics of microblog data. Comput. Graph. **38**, 51–60 (2014)
30. Official Gazette, The Philippines
31. Boquet, Y.: The Philippine Archipelago (2017)
32. Guha-Sapir, D., Vos, F., Below, R., Ponserre, S.: Annual disaster statistical review 2011: the numbers and trends. Centre for Research on the Epidemiology of Disasters (CRED) (2012)
33. Israel, D.C., Briones, R.M.: Impacts of natural disasters on agriculture, food security, and natural resources and environment in the Philippines (No. 2012-36). PIDS discussion paper series (2012)
34. Benavidez, R., Jackson, B., Maxwell, D., Paringit, E.: Improving predictions of the effects of extreme events, land use, and climate change on the hydrology of watersheds in the Philippines. Proc. Int. Assoc. Hydrol. Sci. **373**, 147–151 (2016)

35. Intergovernmental Panel on Climate Change (IPCC)
36. Leichenko, R.: Climate change and urban resilience. Current opinion in environmental sustainability 3(3), 164–168 (2011)
37. Solecki, W., Leichenko, R., O'Brien, K.: Climate change adaptation strategies and disaster risk reduction in cities: connections, contentions, and synergies. Curr. Opin. Environ. Sustain. 3(3), 135–141 (2011)
38. Jones, H.P., Hole, D.G., Zavaleta, E.S.: Harnessing nature to help people adapt to climate change. Nat. Clim. Change 2(7), 504–509 (2012)
39. Neumayer, E., Barthel, F.: Normalizing economic loss from natural disasters: a global analysis. Glob. Environ. Change 21(1), 13–24 (2011)
40. Govind, P., Verchick, R.R.: Natural Disaster and Climate Change. International Environmental Law and the Global South: Comparative Perspectives (2015)
41. Anderson, J., Bausch, C.: Climate change and natural disasters: Scientific evidence of a possible relation between recent natural disasters and climate change. Policy Brief for the EP Environment Committee (IP/A/ENVI/FWC/2005-35) Brief Number 02a (2006)
42. Leng, G., Tang, Q., Rayburg, S.: Climate change impacts on meteorological, agricultural and hydrological droughts in China. Global Planet. Change 126, 23–34 (2015)
43. Department of Interior and Local Government. Climate Change in the Philippines (2011)
44. Garcia, K., Lasco, R., Ines, A., Lyon, B., Pulhin, F.: Predicting geographic distribution and habitat suitability due to climate change of selected threatened forest tree species in the Philippines. Appl. Geogr. 44, 12–22 (2013)
45. Wirtz, A., Kron, W., Löw, P., Steuer, M.: The need for data: natural disasters and the challenges of database management. Nat. Hazards 70(1), 135–157 (2014)
46. Zhang, Z., Wang, P., Chen, Y., Zhang, S., Tao, F., Liu, X.: Spatial pattern and decadal change of agro-meteorological disasters in the main wheat production area of China during 1991–2009. J. Geogr. Sci. 24(3), 387–396 (2014)
47. Alamgir, M., Shahid, S., Hazarika, M.K., Nashrrullah, S., Harun, S.B., Shamsudin, S.: Analysis of meteorological drought pattern during different climatic and cropping seasons in Bangladesh. JAWRA J. Am. Water Res. Assoc. 51(3), 794–806 (2015)
48. Felbermayr, G., Gröschl, J.: Naturally negative: the growth effects of natural disasters. J. Dev. Econ. 111, 92–106 (2014)
49. Parwanto, N.B., Oyama, T.: A statistical analysis and comparison of historical earthquake and tsunami disasters in Japan and Indonesia. Int. J. Disaster Risk Reduct. 7, 122–141 (2014)
50. Li, C.J., Chai, Y.Q., Yang, L.S., Li, H.R.: Spatio-temporal distribution of flood disasters and analysis of influencing factors in Africa. Nat. Hazards 82(1), 721–731 (2016)
51. Kamber, M., Han, J., Pei, J.: Data mining: Concepts and Techniques. Elsevier, Amsterdam (2012)

S/W Engineering and E-Commerce

Improving Software Automation Testing Using Jenkins, and Machine Learning Under Big Data

Ali Stouky$^{(\boxtimes)}$, Btissam Jaoujane, Rachid Daoudi, and Habiba Chaoui

Laboratoire Génie Des Système, L'équipe ADSI, ENSA de Kénitra,
Kenitra, Morocco
ali.stouky@gmail.com, btissam.jaoujane@gmail.com,
rachiddaoudi17@gmail.com, mejhed90@gmail.com

Abstract. Software testing is an essential phase of software development life cycle that ensures quality of the software by fixing bugs which can be done with automated testing to reduce human intervention and to save time and effort consumed in the manual testing. The entire process of testing can be automated easily with the help of automated testing tools. This paper provides a feasibility study for the most commonly used testing tools, we conducted a comparative study of five open source tools to determine their usability and effectiveness. Another point discussed in our paper is the use of machine learning under big data in order to make the system intelligent so that tests lend themselves to automation. We will show how can the combination of all these mentioned technologies can help users to decide which strategy to go for to save both cost and time during testing phases.

Keywords: Big data · Machine learning · Test automation
Software testing tools · Functional testing

1 Introduction

Software testing is the process of evaluating a system to check that it satisfies the specific requirements and identify the differences between expected and actual outcomes. The objective behind software testing process is to identify all the defects existing in a software product [1]. Software testing can be done manually or automatically, manual testing is error prone and a time-consuming process it become a bottleneck in the enterprise field when multiple tests are executed daily [2].

For that firms needs to speed up the testing time, cut costs and reduce data maintenance effort by automating the software testing. Automation Testing is a term that refers to the use of testing tools to cut the need of manual or human intervention, repetitive or redundant tasks, and discover defects manual testing cannot expose.

In software development, developers are very good at delivering code that performs the client's needs and within the agreed timeframe. But that often comes at a cost, the code quality is not in the required standard because tests are not conducted as they have to be, some use cases are missing or sometimes there are software regressions caused

© ICST Institute for Computer Sciences, Social Informatics and Telecommunications Engineering 2018
J. J. Jung et al. (Eds.): BDTA 2017, LNICST 248, pp. 87–96, 2018.
https://doi.org/10.1007/978-3-319-98752-1_10

by files overwritten accidentally by another developer when there's a team of developers involved. Testing is a task that developers take lightly, they rely on others to do it for them and notify them in case there is a bug. Every time a version of the software is delivered, the project manager is not really sure about what to expect.

And here where automated testing comes to picture. In this paper, we will also show how Big Data and machine learning can help us achieve better results.

1.1 Need of Automated Testing in Big Data

The technological developments in big data infrastructure, analytics, and services allow firms to transform themselves into data-driven organizations. Moreover, Big Data helps companies to achieve higher performance like faster problem-solving issues that used to take years to be solved are now taking less time, hence saving the company various dollars and man-hours.

Imagine if we have ten teams in our company each consists of ten developers and each developer works on 10 features per day and runs their test scenarios and generate their reports. In that case we will have 100 scenarios per team and 1000 per day for the whole company, these data are therefore large and the use of big data will allow us to save data processing time.

For that companies of all sizes need a strategy for big data and a plan of how to collect, use, and protect it, every firm needs to build capabilities to leverage big data in order to stay competitive.

2 Machine Learning (ML) Under Big Data

ML is a core research area of artificial intelligence, whose theme is to emulate human learning activities [3]. It studies methods of identifying current and acquiring new knowledge and improving qualities to realize self-perfection.

Machine Learning Under Big Data Can Help Overcome the Challenges in How Can the Workflow Be Automated by Adopting a Developing Trend Described as Follows:

Transfer Learning: The ability of knowledge transferring and transforming between different scenarios is called transfer learning [3]. The traditional machine learning algorithms usually just solve isolated problems. Transfer learning is based on storing knowledge gained while solving one problem and applying it to a different but related problem to improve the learning performance.

As we can see in the Fig. 1 above, traditional machine learning techniques try to learn each task from scratch, while transfer learning techniques try to transfer the knowledge from some previous tasks to a target task when the latter has fewer high-quality training data [4].

In our case, developers usually run their tests scenarios from scratch every time a build was done, which can be time and effort consuming of course. Thus, by adopting this trend the system will be intelligent and if a developer runs a test it will transfer the results and the scenarios to other nodes making them able to profit from those results and to learn from them.

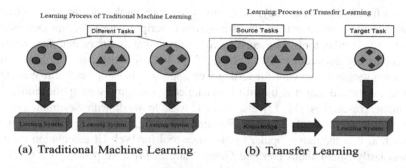

(a) Traditional Machine Learning   (b) Transfer Learning

Fig. 1. Difference between Traditional Machine Learning and Transfer Learning

3 Literature Review

Thousands of developers usually work on a local version of the software on their own machines to avoid any problems that can affect other's work, so they add their modifications and the new features repeatedly, and then merge their new changes with the latest build of the project that is ready to be deployed.

For any version there exists multiple in home local builds of that same version of code at each developer's site and while many developers are coding in the same time, it makes it hard to test every local version that a developer has modified or added, and as we know if the team works daily and most of the time nightly at home or weekly, we are going to have builds that are impossible to test adequately. And thus the question that comes to mind is what framework should we use, or how can we decide which solution is better for the company.

In the past 10 years a lot of researches have been done in this scope, Ramler et al. in [5] discussed the benefits extracted from the automated testing. They presented test automation as one solution to reduce testing costs and proposed a cost based model, to decide on automation strategy.

Kasurinen et al. in [6] examined the industrial applications of the automated testing. Theydeduced that approximately 26% of the test cases are automated in industry, the adoption of such approach is a demanding step in software industry and most of the time test automation is used for quality control and quality assurance.

In [8]: Vahid Garousi et al. presented a multi-vocal review on test automation defining when to automate and what to automate. They gave in details the background of test automation.

And lately Mohamed Monier, Mahmoud Mohamed El-mahdy, focused on the evaluation of automated web testing tools Computer. They provided a feasibility study for some few commercial and open source web testing tools.

In [9] Divya Kumar et al. have tried to identify and classify the tools in which the test automation success depends on the three critical software dimensions of time, cost and quality. The statistics and result of their experiments clearly show the positive effects of test automation on cost, quality and time to market of the software.

While in [1], Yogesh Kumar presented a comparative study on testing tools in order to do the selection of right automated software testing tool, they have discussed four tools that fit into their case study: According to their research Selenium can reduce the cost as it is open source but the efforts involved in scripting for selenium increased by about 15% than other tools. The same for S. Rajeevan that discussed automation testing tools comparing Selenium to Quick Test. As for In [10] Satish have implemented an automation testing framework for testing web applications using selenium WebDriver tool and mentioned that it is useful for dynamically changing web applications.

Another researches [11–13] also have chosen to work with Selenium and that shows that it is commonly used for Functional Testing. To sum up, they mentioned Selenium as it is the most popular open-source tool for Web UI automation testing.

And lastly Sinno Jialin Pan and Qiang Yang in [4] have presented a survey on transfer learning. Beginning from its history, to the relationship between traditional Machine Learning and various transfer learning settings, to its application domains and technologies. They defined what to test how to test it and when in details.

A lot more researches have been done in this context witch highlight on the benefits and the value of test automation. However no work, to the best our knowledge, has been done to show the difference between the tools who exists nowadays that we can use for continuous integration including all test phases and how can we use them and to reduce the cost what are the tools that are open source and can provide a user based model to run functional testing and managing the use of ML under Big data as we discussed earlier.

4 Evaluation of the Study

Automated testing tools can improve the testing effort if requirements are well defined and managed, and the right tool that matches the needs is selected and most important thing is that tests should lend themselves to automation.

In this section we are going to present five test automation tools which are the most used in the past few years, and in the next section we are going to do a detailed comparison between two of these five tools in order to present the tools that are adequate for our approach.

In our case study we analysed the Integrated Development Environment of two of these tools and performed functional testing of a web application developed with PHP5 using Laravel: the PHP Framework for web artisans and both database management systems: MySQL and MongoDB, also for versions management tool we used Bitbucket a web-based hosting service that is similar to GitHub witch primarily uses GIT and it has three deployment models: Cloud, Bitbucket Server and Data Center.

4.1 Apache Jmeter: Load and Performance Tester

It is an Apache Open source load testing tool, written in Java 6+ and supports all platforms. Recently, Apache released the stable version of Jmeter "v2.11" that supports all platforms. Basically, Jmeter is used for load testing and to analysing and measuring the performance of system/application. It is capable to check the performance of the SOAP, LDAP, Message-oriented middleware (MOM) via JMS,

Mail (SMTP(S), POP3(S) and IMAP(S)), Mongo DB (NoSQL), and Native commands or shell scripts. Its strong GUI design helps in fast building of Test Plan and debugging process.

4.2   Citrus Integration Testing

Automated integration tests for message protocols and data formats! HTTP REST, JMS, TCP/IP, SOAP, FTP, SSH, XML, JSON …Citrus is a testing tool that enables the test team to define whole use case tests to be executed fully automated. Incoming and outgoing messages are predefined in the test case. The tester defines a message flow as it is designed for a use case. All surrounding interface partners are simulated regardless their transport protocols (Http, JMS, TCP/IP, SOAP, and many more).

4.3   Selenium

Selenium suite is comprised of four basic components; Selenium IDE, Selenium RC, WebDriver, Selenium Grid. Selenium IDE is Firefox add-on for record-and-playback web application tests. WebDriver directly communicates with the web browser and uses its native compatibility to automate. It is implemented as a Firefox extension, and allows you to record, edit, and debug tests.

4.4   Codeception

Codeception collects and shares best practices and solutions for testing PHP web applications. With a flexible set of included modules tests are easy to write, easy to use and easy to maintain. Codeception encourages developers and QA engineers to concentrate on testing and not on building test suite. Codeception supports major frameworks: Symfony, Silex, Phalcon, Yii, Zend Framework, Lumen, and Laravel.

4.5   Jenkins

An open source automation tool written in Java with plugins built for Continuous Integration purpose. Jenkins triggers a build for every change made in the source code repository for example Git repository. Once the code is built it deploys it on the test server for testing. Concerned teams are constantly notified about build and test re-sults. Finally, Jenkins deploys the build application on the production server.

With Jenkins, organizations can accelerate the software development process through automation (Table 1).

Table 1. Comparison of open source automated testing tools

Features	Tools				
	Selenium	JMeter	Citrus	Codeception	Jenkins
Operating System/Platform	ALL	ALL	ALL	ALL	ALL
Browser Support	ALL Browsers	Google Chrome, IE, Mozilla Firefox, Opera	Google Chrome, IE, Mozilla Firefox, Opera	ALL Browsers	ALL Browsers
Language and Frameworks Support	Php, java, C#, javascript, perl, python, Ruby	Java, NodeJS, PHP, ASP .NET	Java, Xml	Php, Ruby, Python, Java, .Net, Appium, NodeJS, Javascript, Robot Framework	Python, Ruby, Java, Android, C/C++
Ease Of Use	Needs a quite Expertise	No coding is necessary at the basic level	Experience needed	Experience needed	Very easy to use
Software Cost	Open Source	Licence Apache	Licence Apache	Open Source	Open Source
Type Tests	Unit, functional	Performance	Automated integration tests	Acceptance, functional, unit	Unit, Automated integration tests
Script Creation Time	Low	High	High	High	Low
Report Generation	Integration with jenkins can give good reporting & dashbord capabilitie	Graphic, spline, assertion result, tree, statistics	test plan and document test coverage	Several information is provided: execution time, statistics	Checkstyle, PHPMD, PHPCPD, HTMLReport...

5 Analysis of Study

Even if it is known by its Ease of use Jmeter doesn't work well when you have frequent components that you reuse across tests or having different modular tests chained together to form a bigger load tests. It gets harder to do so as you progress with more tests or more levels of testing. The lazy way most beginners would deal with reuse is copy & paste from one area to another or one file to another witch can affect the efficiency of the work.

As for Citrus, it is an integration testing platform for testing live applications deployed in a target environment. In our case we aim to do functional testing as long as continuous integration which cannot be done with Citrus. Also based on the benchmark table and what have been said previously we decided to analyse three of the proposed tools:

5.1   Codeception

First of all, Codeception combine all testing levels. Out of the box you have tools for writing unit, functional, and acceptance tests in a unified framework. Perfect for REST and SOAP API testing and tests can be written in BDD format with Gherkin also it allows multi-request functional tests. In the Table 2 below we present in details its pros and cons for each testing phase.

Table 2. Codeception: pros and cons of functional testing phase

	Pros	Cons
Functional test	• Can be run on any website • Can provide more detailed reports • You can still show this code to managers and clients • Stable enough: only major code changes, or moving to other framework, can break them	• JavaScript and AJAX can't be tested • By emulating the browser you might get more false-positive results • Requires a Framework • Fewer checks can lead to false positive results

5.2   Jenkins

Jenkins can automate the building of software regularly, and trigger tests pulling in the results and failing based on defined criteria. Failing early through build failure lowers the costs, increases confidence in the software produced, and has the potential to morph subjective processes into an aggressive metrics-based process that the de-velopment team feels is unbiased (Table 3).

Table 3. Comparison between "Before and After Jenkins"

Before Jenkins	After Jenkins
The entire source code was built and then tested. Locating and fixing bugs in the event of build and test failure was difficult and time consuming, which in turn slows the software delivery process	Every commit made in the source code is built and tested. So, instead of checking the entire source code developers only need to focus on a particular commit. This leads to frequent new software releases
Developers have to wait for test results	Developers know the test result of every commit made in the source code on the run
The whole process is manual	You only need to commit changes to the source code and Jenkins will automate the rest of the process for you

5.3 Selenium

Because of its many advantages, Selenium finds wide usage for UI, regression, unit and acceptance testing. But even selenium is the most popular tool it stills not a complete, comprehensive solution to fully automating the testing of web applications. It requires third-party frameworks, language bindings and another configuration to be truly effective.

6 Proposed Solution

It is obvious from the above stated cons and issues that not only the software delivery process became slow but the quality of software also went down. This leads to customer dissatisfaction and no tool will manage the whole process of Testing along with continuous integration and continuous deployment. So to overcome such a chaos there was a drastic need for a system to exist where developers can continuously trigger a build and test for every change made in the source code.

On top of Jenkins is an open source technology, so the code is open to review and has no licensing costs. And it is a master slave topology that distributes the build and testing effort over slave servers with the results automatically accumulated on the master. This topology ensures a scalable, responsive, and stable environment.

Jenkins is the most mature CI tool available and in the following diagram we are going to illustrate how Continuous Integration with Jenkins overcame the above shortcomings along with the use of Selenium framework to guarantee perfection needed in the entire workflow (Fig. 2).

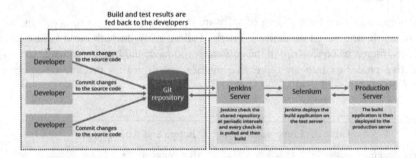

Fig. 2. Generic flow diagram of Continuous Integration with Jenkins

- First, when a developer commits the code to the source code repository the Jenkins server checks the repository at regular intervals for changes.
- Soon after a commit occurs, the Jenkins server detects the changes that have occurred in the source code repository. Jenkins will pull those changes and will start preparing a new build, If the build fails, then the concerned team will be notified else If built is successful, then Jenkins deploys the built in the test server (Selenium) to run tests and generate reports properly.

- After testing phase, Jenkins generates a feedback and then notifies the developers about the build and test results. It will continue to check the source code repository for changes made in the source code and the whole process keeps on repeating.

The other important key point as we discussed earlier is the combination of this approach of testing with the use of Big data to have a solid architecture so that machine learning can have a basic source of information or outcomes to use them as incomes, and hence to learn from experiences, findings and mistakes of those tests who have previously been passed.

The Learning process from those results transforms the system to become intelligent so the testing phases work fluently and in a predictive way too. Machine Learning needs to develop and progress to change big data into actionable insight. On one side, big data provides rich information for Machine Learning algorithms to pick up underlying patterns and to build predictive models; and on the other, traditional Machine Learning algorithms face crucial challenges like scalability to release the true value of big data.

7 Conclusion

Software testing tool can be selected based on Application needed to be tested, budget, usage and the efficiency required. Our comparative study is helping to select the suitable tools based on multiple criterion. It presents each tool with features that are in the same and different degree with other tools mentioned and how each one behaves against others tools' characteristics.

Producing quality requires a great attention to details. Jenkins can pay attention to many of the details and shout loudly when violations occur.

In this paper, we have also shown how the adoption of big data can help a company to achieve better results and high performance in order to reduce time and cost and how firms nowadays must go for big data technologies, moreover optimizing tests execution time and effort by using Machine learning algorithms to automate the whole process and to make the system intelligent. Our future work will encounter more tools and more technologies also that will help in building a user based smart-model.

References

1. Kumar, Y.: Comparative study of automated testing tools : selenium, SoapUI, HP unified functional testing and test complete. 2(9), 42–48 (2015)
2. De Castro, A.M.F.V., Macedo, G.A., Dias-neto, A.C.: Extension of Selenium RC Tool to Perform Automated Testing with Databases in Web Applications, pp. 125–131 (2013)
3. Shi, C., Wu, C., Han, X., Xie, Y., Li, Z.: Machine learning under Big Data, no. Emim, pp. 301–305 (2016)
4. Pan, S.J., Fellow, Q.Y.: A survey on transfer learning, pp. 1–15 (2009)
5. Ramler, R., Wolfmaier, K.: Economic perspectives in test automation: balancing automated and manual testing with opportunity cost, pp. 85–91 (2006)

6. Kasurinen, J., Taipale, O., Smolander, K.: Software test automation in practice : empirical observations, vol. 2010 (2010)
7. Amannejad, Y., Garousi, V., Irving, R., Sahaf, Z.: A Search-based Approach for Cost-Effective Software Test Automation Decision Support and an Industrial Case Study (2014)
8. Garousi, V., Mäntylä, M.V.: When and what to automate in software testing ? A multi-vocal literature review, vol. **76**, 92–117 (2016)
9. Kumar, D., Mishra, K.K.: The Impacts of test automation on software' s cost, quality and time to market. Procedia Comput. Sci. **79**, 8–15 (2016)
10. Gojare, S., Joshi, R., Gaigaware, D.: Analysis and design of selenium webdriver automation testing framework. Procedia Comput. Sci. **50**, 341–346 (2015)
11. Rajeevan, S., Sathiyan, B.: Comparative study of automated testing tools : selenium and quick test professional, **3**(7), 7354–7357 (2014)
12. Angmo, M.R., Sharma, M.: International Journal of Emerging Technologies in Computational and Applied Sciences (IJETCAS) Selenium Tool : a web based automation testing framework, pp. 351–355 (2014)
13. Singla, S.: Selenium keyword driven automation testing framework, **4**(6), 125–129 (2014)

Mining IT Job-Order Services: Basis for Policy Formulation & IT Resource Allocation

Alvin Jason A. Virata[1(✉)] and Jasmin D. Niguidula[2]

[1] St. Dominic College of Asia, Talaba IV, City of Bacoor, Cavite, Philippines
ajavirata.7772011@gmail.com
[2] Technological Institute of the Philippines, 1328 Arlegui St.,
Quiapo, Manila, Philippines
jasniguidula@yahoo.com

Abstract. As the computer hardware have been globally accepted in the areas of industry and education, the IT Job- Order Service Delivery are now becoming very essential part in the resources management. With fast-paced technology development, decision makers require a valid reporting IT resources management [1]. IT resources were expected to be fully maintained to ensure the long-term life span of the computer hardware and devices. Further, having an IT Resource Management System or Software will not be sufficient since maintenance and repairs may probably be a burden to the budget allocation when there is a demand for IT infrastructure development or innovation [2]. Determining the instances of IT Job Order Services, its trends of repairs and services conducted has a significant implication to the decision makers. Using a predictive analytics of the data sets in the IT Job-Order services, could be a basis for revising the policy and IT resource allocation.

Keywords: Predictive analytics · Resource management
Policy improvements

1 Introduction

The prevalence of technology innovation has expanded considerably that brought the management philosophy to become a technology-driven oriented towards the improvement of goods or services [3]. Subsequently, business organization have worked persistently to think of new techniques for interfacing with clients. Key indicator to this procedure has been the improvement of e- ticketing (job-order) process/request. [4] Facilitating ticketing system in IT Job- order services turns to be one of the most important undertakings in dealing with the business operations. Ticket analytics is necessary to distinguish peculiarities to identify uncommon patterns in the operations [5].

Forecasting or conducting predictive analytics may be measured based on analysis of the extracted data taken from the gathered information that can be utilized to anticipate future patterns and conduct new technique designs. An Application Management System (AMS) Analytics framework gives propelled competency analysis on

© ICST Institute for Computer Sciences, Social Informatics and Telecommunications Engineering 2018
J. J. Jung et al. (Eds.): BDTA 2017, LNICST 248, pp. 97–106, 2018.
https://doi.org/10.1007/978-3-319-98752-1_11

learning management, asset administration and allocation of operational resources as well as operational forecast analysis model [6].

Minimizing the personnel efforts in administration of IT service management requires mechanism to expedite the routine maintenance method. There are three (3) research directions that are being recognized or considered to support the IT Service Administration Management: (1) Automatically find out the manifestation of symptoms when facilitating IT Job Service; (2) Logically assess the pattern of events; (3) Identify transient dependencies within the performance statistical data. Furthermore, executions of data framework, could be viably addressed with data-driven result [7].

Generally, the main focus of this study is to predict the best practice processes in implementing the IT Infrastructure Support Services. Specifically, it gears to: (1) identify the frequent request for IT Job-Order Service that was performed to each department, (2) perform predictive analytics on IT Job-Order trends, and (3) recommend appropriate practice in allocating budget in the acquisition of IT equipment and devices.

Moreover, the study is significant to every decision maker of any business organization to formulate and implement policies in allocating IT resources effectively on increasing cost-effectiveness.

2 Related Works

"Application Management Services (AMS)" is a kind of IT outsourcing administration offering that utilizes coordinated procedures, approaches, methodology and norms to deal with the customers' request record. Ordinarily, it provides administration help desk and application support. An application management service has structured and unstructured information. The structured information were classified when specific data were identified. On the other hand, unstructured information is literary depiction of the issue and the documentation of the action taken. The expansive objectives of ticket investigation incorporate with (1) evaluation of ticket operational workload volume; (2) evaluation of ticket action taken from the time AMS we assigned a proof of operational proficiency; (3) proof of potential issues that arise during the service delivery [8].

On the other hand, IT Management Service (ITSM) was considered as one of the IT contribution to customer's business operation. ITSM is also unique in relation to the customer technology-centered methodologies to IT management and business operation. [9]. Moreover, IT Service management process were also known as "Incident Management" [8]. The goal of IM System is to restore the regular service operation as immediate possible subsequently, it guaranteed that those best workable levels of service administration were practiced. [10].

Predictive analytics is defined as part of analysis that focus on the extraction of data from a gathered information that will be used to identify the future patterns and behavioral trends. It is dependent on the connections between variables and from the past predicted variables [4].

In the merchants groups, there is a particular framework the utilize the "ITIL as their thrust: IBM's Process Reference Model for IT (PRM-IT), Hewlett Packard's

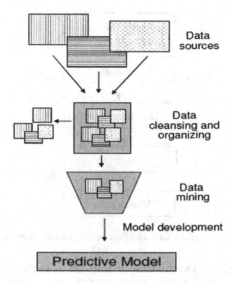

Fig. 1. IBM predictive analytics framework

ITSM Reference Model, and Microsoft's Operating Framework (MOF)." All ofthese frameworks have their own unique approaches to the utilization and execution of ITSM, assisted by an organization for the software proprietary [7] (Fig. 2).

3 Methodology

Investing in IT infrastructure services is highly beneficial and is cost-effective in the long haul [11]. While completely eliminating risk is impossible to achieve, it can help in having increased reliability to significantly reduce risks. To address the purpose of the study in data mining the IT Job-order service, the following methodology were considered:

1. Identify the appropriate tools in data mining the IT Job-Order service. Gather the IT job-order service record for one year.
2. Identify the dataset to be used for predictive analytics
3. Generate a visualize output in terms of:
   a. Hardware category versus request status
   b. Requesting department versus request status
   c. Request status versus hardware category
   d. Frequency of IT service request for department.
4. Identify how IT technical support and services were effectively facilitated through technical support and services.
5. Cluster similar/recurring concerns or problems encountered
6. Cluster the hardware category between old and new units/devices

Fig. 2. Mining data of IT job-order service flowchart

The researcher used the Orange Canvas software as tools to facilitate data mining predict analytics on the IT Job-order Services [12]. Orange canvas is an open source machine learning for data visualization. It provides interactive data exploration for a rapid analysis and visualizations of data sets that provided a fast data prototyping and analysis. Further, a data set was gathered based on the one-year IT Job-Order record of the ICT Department. The following data sets that was used for data mining predictive analytics are (1) Department, (2) request, (3) Job Order Number Series (4) Hardware classification (5) request category, (6) problems encountered, (7) Action taken (8) date of Job-order service.

4 Results and Discussions

This section discusses the data mining results based on the different parameters from the visualization analysis tool using the Orange Canvas. The results of the visualization shows the comparison of different entities which relates to the, frequent IT Job-Order service request, Common IT Job-order Service delivered, and identifying the hardware status/condition that are requested for IT services.

Figure 3, Shows that most of the client's IT Job-Order Service requests were referring to the System Units. Result reveals that 60% of the data refers to the request of old unit. In the same vein, the remaining 40% reveals that it is still allocated for the IT Job-order service request of system units that are new. One of the most intrinsic responsibilities of a Management Information System Office is to maintain the quality of the computer services that relates the company's hardware and software assets. It is a must to detect technical issues early, before they become problems. The visualization shows that repairing of old system units is a normal maintenance activity, it is alarming that concurrently, requests for the IT job order service for new system units is almost as high as the result for the maintenance of the old units. This goes to show that the administration should develop a policy on proper handling of computer units to reduce the problems on the increase of new system units maintenance.

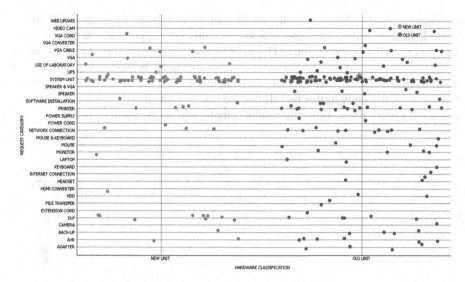

Fig. 3. Graphical representation of the IT job-order service request

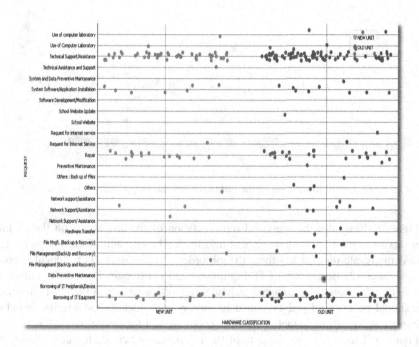

Fig. 4. Visualization of the request classification on the IT job-order service request

Figure 4, Shows the request classification of the IT job-order request based on the hardware classification of old and new system units that supports the result of Fig. 1. The data reveals that the top three IT job-order request are technical support (such as troubleshooting and preventive maintenance), borrowing IT peripherals/equipment, and repair (minor and major replacement of computer parts). As related to the hardware classification, the study further reveals that the new system units repairs are also as high as the old system units. Thus, the results compliments to the findings in Fig. 1.

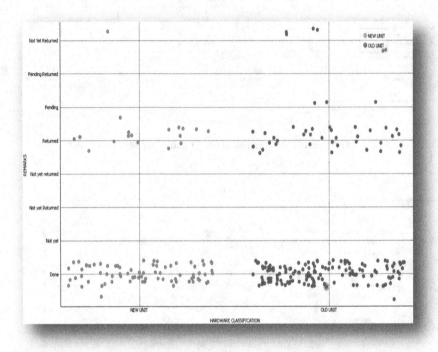

Fig. 5. Graphical illustration of the task status on the IT job-order service request

Figure 5, Illustrates the graphical representation of the task status on the IT job-order service request to the management information system unit of the company. The result demonstrate that the top three (3) job order services were (1) technical support and assistance (2) Borrowing of IT Equipment and (3) Repair. Hence, the outcomes indicate that the issue on repairs and technical support/assistance of the newly acquired units should be assessed and evaluated to appropriately identify the specific brand of equipment and peripherals that should be recommended for future purchases.

Figure 6, Shows the visualization of the IT Job-order Service Request among the different departments inside the organization. The result shows that the acquisition of request is fairly distributed among the departments. Thus, the finding reveals that the policy on IT hardware maintenance should be revisited for update and improvement to provide policy on priority of acquisition of system units and other peripherals.

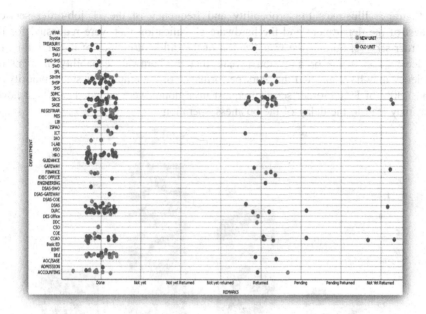

Fig. 6. Visualization on the IT job-order service request among the department and task status

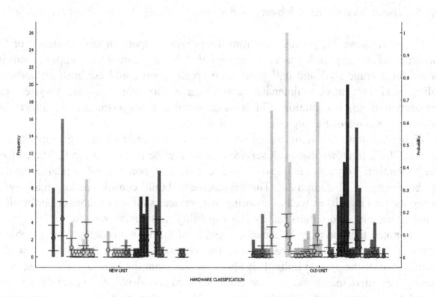

Fig. 7. Visualization of the different entities of the IT job-order service request in the organization

Figure 7, Illustrates the probability and frequency of the IT job-order service request of the department and the hardware classification that were requested for IT job-order service. Further the visualization of the different entities of the IT job-order service request in the organization reveals that the hardware capability in terms of efficiency may have affected the work performance of the employees since majority or almost 30% of IT job-order service request in a month were attended in which however, may lead to the delay of the business operation.

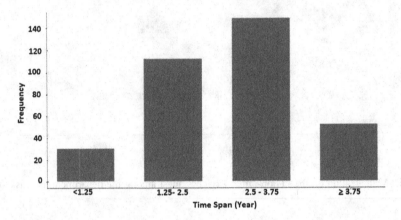

Fig. 8. Visualization on the IT job-order service request among the department and task status

Figure 8, However, proves that from the previous report on the frequency of IT Job-order service request it was very alarming that the life span of the computer units is mostly an average of 2 and half years. The organization could establish an updated policy of IT preventive maintenance and resource allocation to create a systematic approach in the implementation of IT preventive maintenance procedure as it affects the business processes of the organization as a whole.

Figure 9, shows the visualization of the job order service according to month for the year 2017. It shows that peak service months are the months of April, May, June, and November. Top services provided are technical support, website update, repairs, and borrowing of IT equipment. The management should consider adding manpower during the said months as well as ensuring that services staff are capable of performing technical support, and repairs, as well as capability in updating website.

It is noted that the overall frequency results of the IT job-order service graphical representation show that the most common request for technical services was the maintenance of the system units. It is also important that the existing process in conducting preventive maintenance of the computer hardware deployed to each department should be reviewed and studied to create a better solution in technical assistance. Moreover, the cause distribution can be referred to: (1) lack of personnel technical knowledge, (2) the efficiency of the technical manpower conducting hardware assistance, (3) the proper deployment of system units. These three (3) areas of distribution should also be taken for action to contribute in strengthening policy implementation.

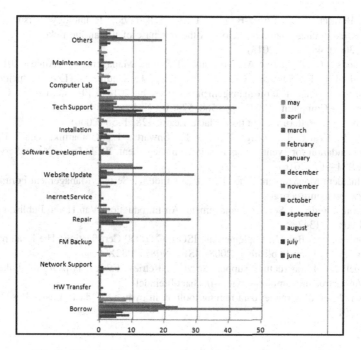

Fig. 9. Visualization of job order service according to month

5 Conclusions

Information Technology in business organization has become an important tool to improve business process. The study deals on the predictive analytics on the IT Job-Order Services. The data reveals that the number of the job-order services request is equally distributed among the departments. However, since maintenance services of new hardware peripherals is as high as the old ones it is also important to check, validate and assess the kind of hardware peripherals and equipment allocated for each department. A need to compare the service level of maintenance outcome between branded and cloned computers should be given a priority. Thus, it is concluded that the organization must establish an updated policy of IT preventive maintenance and resource allocation to create a systematic approach in the implementation of IT preventive maintenance procedure as it affects the business processes of the organization as a whole.

References

1. Markgraf, B.: The Role of Management Information Systems in Decision-Making. Retrieved 2017, Chron. http://smallbusiness.chron.com/role-management-information-systems-decisionmaking-63454.html
2. Mishra, R.K.: Role of Information Technology in supply chain management (2014)

3. Lopez-Bonilla, J.M., Lopez-Bonilla, L.M.: Self-service technology versus traditional service: examining cognitive factors in the purchase of the airline ticket. J. Travel. Tour. Mark. **30**(5), 497–513 (2013)
4. Ed Woods, I.C.: Predictive Analytics and IT Service Management, 8 August 2012
5. Li, T.H., Liu, R., Sukaviriya, N., Li, Y., Yang, J., Sandin, M., Lee, J.: Incident ticket analytics for IT application management services. In: 2014 IEEE International Conference on Services Computing (SCC), pp. 568–574. IEEE, June 2014
6. Bishop, C.M.: Pattern recognition. Mach. Learn. **128**, 1–58 (2006)
7. Zhou, W., Tang, L., Zeng, C., Li, T., Shwartz, L., Grabarnik, G.Y.: Resolution recommendation for event tickets in service management. IEEE Trans. Netw. Serv. Manag. **13**(4), 954–967 (2016)
8. ITIL Incident Management: The ITIL Open Guide, IT Service Management Forum – USA. http://www.itsmfusa.org/
9. Van Bon, J. (ed.): IT Service Management: An Introduction. Van Haren Publishing (2002). ISBN 9080671347
10. Clifford, D.; van Bon, J.: Implementing ISO/IEC 20000 Certification: The Roadmap. ITSM Library. Van Haren Publishing (2008). ISBN 908753082X
11. Breindel, M.: Infrastructure Support Services. Retrieved 2017. https://www.linkedin.com/pulse/infrastructure-support-services-michael-breindel
12. Demsar, J., et al.: Orange: data mining toolbox in python. J. Mach. Learn. Res. **14**, 2349–2353 (2013)

eCommerce Sales Attrition: A Business Intelligence Visualization

Leah G. Rodriguez, Jonathan M. Caballero[(⊠)], Jasmin D. Niguidula,
Dawn Iris Calibo, and Christopher A. Rodriguez

Technological Institute of the Philippines (T.I.P.), Manila, Philippines
mam_leah@yahoo.com.ph, jonathanmcaballero@gmail.com,
jasminniguidula@gmail.com, dawniris_19@yahoo.com,
thopsar7@yahoo.com

Abstract. As technology advances, electronic commercialization makes a more significant impact on business as it offers versatility and convenience. In fact, organizations have made a leap to adopt related knowledge frameworks in e-commerce computing with the combination of business intelligence to remain in front of patterns and future demands. This is manifested by a significant number of organizations of all sizes, in all enterprises, all around the globe by actualizing and using Strategic Business Intelligence. This paper deals with the development of an Online Product/Service Catalogs with Business Intelligence application for a shipping and aircraft business that leads to the increase of efficiency and profits through timely and informed decision making. Using descriptive and developmental instrument, system development model and data visualization tools, the researcher evaluate the website's sales attrition after having developed its mobile application version and deploying the system online. The results indicate the usefulness of the system in managing materials, delivering quality services to the customers and at the same time helps businesses to increase competitiveness and business profitability.

Keywords: Sales patronage · Business · Data mining · Mining visualization
Online shopping

1 Introduction

In the world business, profit indispensable part to its existence. This gives importance to keep track of the company's sales attrition rate to measure its impact on the business profit in particular and to the whole business existence as a whole. As a matter of fact, business establishments progressively need to carry on like multichannel retail traders by concentrating on approaches to venture into new markets, pick up clients, keep them and boost the productivity of every transaction. This paves the way to the existence of electronic commercialization (e-commerce).

E-commerce alludes to the way toward purchasing or offering items or administrations over the Internet. Web-based shopping is ending up plainly progressively well known in light of speed and convenience for clients [1]. This technology enables marketers to widen their horizon and keep their competitiveness with other business entities in the similar market field.

© ICST Institute for Computer Sciences, Social Informatics and Telecommunications Engineering 2018
J. J. Jung et al. (Eds.): BDTA 2017, LNICST 248, pp. 107–112, 2018.
https://doi.org/10.1007/978-3-319-98752-1_12

For those that venture into e-commercialization, it gives an advantage for business companies such as shipping and aircraft services to venture into the world of online retailing. Aside from embracing the technology where customers can avail its services online, the system is equipped with business analytics that measures customer's interaction with the website. These consumers activities are then stored in the system's database and later be processed to visualize important information that could help the company generate the greater picture of the website's current and future trends.

This study deals with the e-commerce sales attrition visualization using business intelligence tools through consumers' interaction in the said shipping aircraft services website.

2   Related Works

Organizations around the globe have grasped web based business in their organizations and have procured benefits thereof. Presently, in web based business, numbers of clients from different social gathering increments through overall promoting of the items. They give distinct advantages to customers as they offer articles and administrations where consumers can browse [2].

Web-based shopping is a necessary plan of action in e-commerce. It is the procedure whereby buyers specifically purchase products or administrations from a merchant continuously, without a mediator benefit, over the Internet. Additionally, it is used as a medium for correspondence to increment or enhance in esteem, quality and appeal of conveying client advantages and better fulfilment, that is the reason web-based shopping is more accommodation and step by step expanding its popularity [3].

Moreover, a huge number of organizations are actualizing and using business insight (BI). It will likely upgrade the importance and estimation of reactions to the choice procedure. Without bi, an organization runs a danger of settling on basic choices in view of deficient and off base data. BI, when effectively thought out and appropriately executed, enables all clients to settle on educated decisions and choices extremely time, in each circumstance [4]. Also, data gathered from a capable bi makes representatives more gainful, providers more productive, and clients more loyal.

With the business intelligence at hand, companies can gather relevant data from the system and out of its visualization, information is generated, and this includes attrition rate.

In the work of Sessoms (2017), he defines the word "attrition" which is commonly used in the in human resources may also be employed in sales. As a matter of fact, sales attrition may refer to the consumer loss and retention. This is directly influenced by different entities such as profit and growth. The importance of keeping track of customers helps the business to improve customer's communication, update marketing strategies, and augment sales and wider market sales. He further explains that silent attrition refers to the loss of clients who leave without communicating their discontent with a business practice or client benefit issue. These clients may react with direct disappointment on client studies. For example, a medical coverage client may feel disappointed with endeavors to achieve the client benefit office and imperfection to another insurance agency. This quiet wearing down happens because the organization is ignorant or inert to the client's misery and does not have a chance to keep the client from clearing out. The client is run with no clarification [5].

In the visualization, study utilizes *Cohort Analysis* to breakdown data structures through the use of algorithm. Cohort analysis is a subset of behavioral investigation that takes the information from a given eCommerce stage, web application, or web based diversion and as opposed to taking a gander at all clients as one unit, it breaks them into related gatherings for examination. These related gatherings, or accomplices, more often than not share basic qualities or encounters inside a characterized time-traverse. Further, it is an apparatus to quantify client engagement after some time. It knows whether client engagement is really showing signs of improvement after some time or just seems to enhance in view of development [6].

With such analysis, the sales attrition rate of a shipping and aircraft company through visualization was realized.

3   Methodology

The research utilizes the experimental method in the processing of data for an aircraft and shipping online shop. In this case, the researchers used a visualization tool incorporated as business intelligence in the said website to identify the sales attrition rate using the data within a year. The data are then presented using the following parameters (Table 1).

Table 1. For mining the data, the study uses the sales attrition, user's statistics, and sales patronage as the basis for the sales visualization.

Parameters used for the sales attrition visualization
Sales Attrition
User's Statistics
Sales Patronage

Furthermore, the study uses the Statistical Analysis System (SAS) for Visual Analytics framework as a basis for projecting actual data of the website and predicting future sales. This framework enables the company to gain insight from all of the data accumulated in the database (Fig. 1).

4   Results and Discussion

This part of the study shows the visualization results based on the different parameters from the business intelligence tool (Fig. 2).

This implies that a great number of customers are inclined to transact in online shopping. This is supported by Katawetawaraks (2011), [8] the study illustrates the enthusiasm of online consumers on what keeps them motivated to online shopping. It is found out that there are a number of reasons why people shop online such as convenience

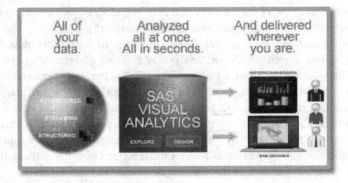

Fig. 1. SAS Visual Analytics Framework is used in the study [*SAS® Visual Analytics*]. The system clarifies that clients can comprehend purchasing designs and distinguishes patterns, which prompts better administration and consumer loyalty. With SAS Visual Analytics, what was made can be distributed and gotten to utilizing tools so managers can see and connect with basic data and basic leadership information effectively [7].

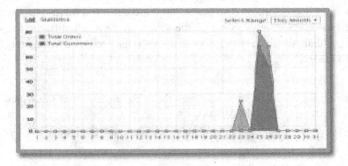

Fig. 2. In this figure, the result shows the graphical representation of the user statistics of the customers who made the website views, reservations and orders within a period of one month. The finding reveals that for the total of 80 customers who views the website, 70 of them avails the online shop's services through reservation of product. It further demonstrates that a 27% of the customers who view the websites completed the close-out transaction since they have completed the order process.

since consumers can buy the same product at a lower price by comparing different websites at the same time; avoid pressure when having a face-to-face interaction with salespeople and even avoid the traffic jam in going to shopping centers (Fig. 3).

In the study of Sultan (2001), it shows that there are some factors which influence online shoppers to transact business on the web and this includes their consideration to prices, discounts, and feedbacks from previous customers and quality of product and services offered [9]. This results then support the findings in the above figure since it is noted that many users are into window shopping and this indicates that they are into viewing and comparing products from another website such that they tend not to buy at once (Fig. 4).

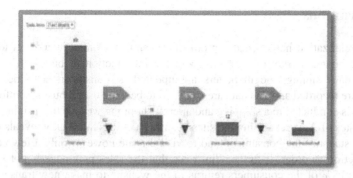

Fig. 3. In this figure, it shows the breakdown of user's statistics that uses the website to shop online. The data reveals that out of 80 users in a month, 18 of them viewed the items, 12 added the items to the cart and 7 of them checked out the items.

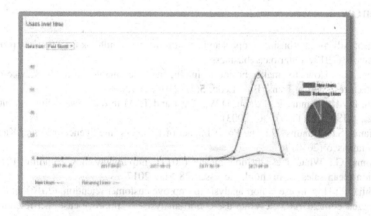

Fig. 4. The figure illustrates the sale patronage of the consumers of the shipping and aircraft online shop based on a one-month data. The finding reveals that out of 80 shoppers 15% are considered to be returnees while 85% of them are new.

Despite the fact that the advance of innovation utilizing e-commercialization develops throughout the years regardless it pauses for a minute for customers to acknowledge it and receive in their everyday life. The absence of buyers' acknowledgment towards new innovation remains a test for the online advertisers. This is clarified in the investigation of Cheema (2013) [10] that the web shop ought to require more endeavors to speak and instruct their objective market about the advantages of the web based shopping on the off chance that they mean to expand the activity on their online shop with the end goal that of a delivery and air ship site.

5 Conclusions

E-commercialization has opened a promising avenue for online marketers to capture wider and bigger market online. And with the introduction of business intelligence tools for data mining, the marketers are updated as consumers activities in their websites are recorded and this data are analyzed to become useful business information. The study is conducted to a shipping and aircraft shopping website that originates in the United States of America. Using the sites business analytics, the study reveals based on the user's statistics the consumers tend to visit the site however only a few completes the transaction to order. It further illustrates that the sales patronage is not very high since only 15% of the consumers returns to the website to make new transactions.

Thus, the study implies that online shopping as a supplement marketing tool has a positive advantage due to its contribution to the sales as it draws a wider range of market by capturing online shoppers.

References

1. Ontario: Advanced e-business topics and an introductory handbook how you can profit from e-business (2013). ontario.ca/ebusiness
2. Panian, Z.: How to make business intelligence actionable through service-oriented architectures. Wseas Trans. Bus. Econ. **5**(5), 210–221 (2008)
3. Gefen, D., Karahanna, E., Straub, D.W.: Trust and TAM in online shopping: an integrated model. MIS Q. **27**(1), 51–90 (2003)
4. Williams, S., Williams, N.: The Profit Impact of Business Intelligence. Morgan Kaufmann, San Francisco (2010)
5. Sessoms, G.: What does attrition mean? (2017). http://www.ehow.com/info_12106271_attrition-mean-sales- terms.html. Accessed 28 May 2017
6. Makhija, P.: How to use cohort analysis to improve customer retention, (2016). http://www.kdnuggets.com/2016/05/clevertap-use-cohort-analysis- improve-customer-retention.html. Accessed 30 May 2017
7. Zhan, L., et al.: Visual analytics for the big data era—a comparative review of state-of-the-art commercial systems. In: 2012 IEEE Conference on Visual Analytics Science and Technology (vast), pp. 173–182. IEEE, October 2012
8. Katawetawaraks, C., et al.: Online shopper behavior: influences of online shopping decision. Asian J. Bus. Res. (2011)
9. Sultan, et al.: Consumers' Attitude towards Online Shopping Factors influencing Gotland consumers to shop online (2011)
10. Cheema, U., et al.: The trend of online shopping in 21st century: impact of enjoyment in tam model. Asian J. Empir. Res. **3**(2), 131–141 (2013)

Social Media and Health Care

Keyboard and Mouse: Tools in Identifying Emotions During Computer Activities

Jheanel Estrada[(⊠)], Jomar Buhia, Albert Guevarra,
and Marvin Rick Forcado

Graduate Programs, Technological Institute of the Philippines,
1338 Arlegui St., Quiapo, Manila, Philippines
jheanelestrada29@gmail.com, buhiajomar@gmail.com,
abetguevarra, coolviper01@gmail.com

Abstract. Emotion recognition is a field of study that is much investigated by researchers because it is an important piece in the development of intelligent system. Researchers used different techniques to capture the emotional response of the people based on the situation and design patterns to see its effects and relevance. In this paper, the researchers investigated the emotional patterns of the students towards programming using input devices such as keyboard and mouse. The use of keyboard and mouse provided an affordable and non-intrusive method of collecting data. The keyboard and mouse data collected were mapped by an expert on facial expressions to emotions captured on video at the same the students were answering the programming problem. From these annotations, the classes of emotions were derived. Rapid miner was then used to identify the pattern of keyboard and mouse strokes that corresponds to different emotions. The accuracy of the results were 70.25% and with a Kappa of 0.612%.

Keywords: Affective computing · Self- assessment · Manikin
Gradient-Boosted Trees

1 Introduction

Emotions play a significant part in our daily lives. These emotions range from feelings of happiness, anger, sadness, etc. Since it is natural for human beings to seek happiness, these feelings impact our overall being [1, 2].

Emotions also play a major role not only in decision -making but also in enhancing learning. Studies on human to human and human to computer intercommunication show that for effective interaction to happen, emotions must be given due importance. By including emotions in human-machine interaction, machine would be capable of detecting the emotional condition of the user and make appropriate adjustments [3]. Affective computing is a discipline that that has user emotions during computer interaction as its focal point [4] and whose primary objective is to develop systems that can recognize and respond to these emotions [5].

Developing a system that recognizes user emotion is not easy as it involves carrying out numerous tasks such as gathering and representing data and system training.

© ICST Institute for Computer Sciences, Social Informatics and Telecommunications Engineering 2018
J. J. Jung et al. (Eds.): BDTA 2017, LNICST 248, pp. 115–123, 2018.
https://doi.org/10.1007/978-3-319-98752-1_13

Nevertheless, many methods for recognizing emotions of computer users have been developed. Some of these are based on analyses of keyboard and mouse movements [6].

Studies by Salmeron-Majadas, et al. [7] and Grimes, et al. [8] used keyboard and mouse interactions along with the Self-Assessment Manikin (SAM) because it provides an easy to understand way of describing emotions [9, 10]. SAM has been an effective tool in measuring responses to various stimuli such as pictures, images, sounds, pain, etc. [11–13].

Most studies on emotion recognition require some sort of special equipment to detect emotion. The current study aims to develop an emotion recognition system by collecting keyboard and mouse data from computer programming students in a non - intrusive manner. The collected data were used, together with the students' answers to SAM questionnaires, as training set and test set for RapidMiner to generate patterns to figure out the emotions of the user. This method provides two main advantages: (1) it does not require any expensive, specialized equipment and (2) it is non -intrusive as it does not meddle with the user's normal computer activities.

2 Related Literature and Studies

Majadas and Santos [14] listed that the relationship between mouse and keyboard use and the user's affective state indicators could be valuable in detecting affective states in non-interfering and inexpensive manner. Ball and Breese [15] further stated in their study that high levels of activity with other applications (typing and mouse clicking rates) also are indicative of high task orientation. Thus, observation of such behaviors should be interpreted as evidence that the user is highly oriented towards task completion.

Bixler and D'Mello, for their part, stated that detecting affect plays an important part in creating applications that converse with and aid users in completing their task. Their study used analysis of keystroke to create affect detectors. This approach requires no additional hardware and presents an effective and unobtrusive method of data collection [16].

The importance of emotion detection is further supported by Zimmerman et al. [17], which stated that in intercommunication between human and computers, expressed emotions have a big role and most people interact with the technology naturally and socially. Importance in HCI research is increasing and the interaction system for emotional intelligence is used to express and respond to emotions of humans. The study is centered on another method of behavior using standard mouse and keyboard devices for measuring user affect and extracts parameters from motor-behavior using log-files of keyboard and mouse actions.

The study considered one of the methodology used for getting information from keyboard-stroke to identify if the emotional state is positive, negative, emotionless or neutral state as the user interacts with the computer. It includes the rate of mouse clicks, the rate of keystrokes as well as the keystrokes average duration as the analyzing parameters.

In the study, Zimmerman, et al., were able to know the outcome of induced effects based on the parameters of motor behavior when the user is using the computers including film clips showing, and doing online shopping. The ninety-six (96) participants were tested using the different mood states as independent variable. The created procedure was used and conducted at the laboratory utilizing the use of sensors and bands that measures respiratory attached to the participants. The data questionnaires displayed on the computer were answered, as well as self-assessment questionnaires, by the participants. The mouse and keyboard actions from the log - files are then analyzed [17].

Another research conducted by Rodrigues, et al., states that when a student attends an e-learning course there is a poorer interaction between student and teacher. There is no verbal interaction and it is difficult for the teacher to assess the feelings and attitudes of the students. The study developed a tool that will estimate the stress level of users without meddling with the user's normal computer activity. Developing dynamic stress estimation model is the main goal of this research. For this research, data collected are click duration and accuracy, amount of movement, mouse movement, mouse clicks and keyboard strokes. It tested 10 programming students to validate the possibility of user's stress detection and analysis of the mouse and keyboard use. It is concluded that stress has significant effect in learning capacities and learning speed [18].

3   Methodology

This study's aim is the detection and identification of the emotions based on the use of non-invasive tools such as the ordinary mouse and keyboard. As shown on Fig. 1(a) the Flowchart for Emotional Classification based on Shukla and Solanki [19]. Data collection is done by performing feature selection and extraction and data labeling follow based on the questionnaire, fuzzy vector and label has been applied for the training of the data and from that classification has been identified. Using data pre-processing and applying the fuzzy vector, the research can display the emotional condition of the user. On the other hand, Fig. 1(b) shows the flowchart of what this research paper is trying to solve.

3.1   Data Gathering Tool

A data gathering prototype program was designed and implemented as a tool used to collect the data needed. The data gathering prototype has a survey question with Self-Assessment Manikin (SAM) images that requires the students to select [19]. The program records the keyboard keystrokes together with mouse clicks and movements then save individually in the .xls (mouse and keyboard) and .txt as the backup files.

The students have also been provided with programming questions for them to answer using Java programming language.

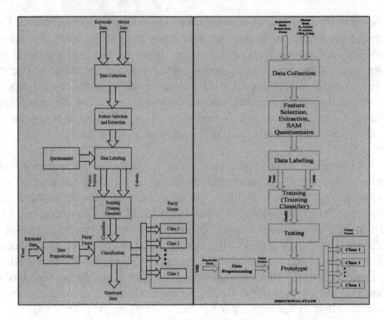

Fig. 1. (a) Flowchart for emotional classification [19] (b) Flowchart for emotional recognition

3.2 Data Gathering Procedures

This research intends to identify and analyze the emotional condition of the computer user. During the data gathering the following actions were conducted by the researchers: (1) randomly selected six (6) participants (three (3) males and three (3) females) who have programming course from a select academic institution. Informed consent form was filled up by the participants. (2) Orientation was done about the assessment that the participants need to answer but with no bias on what data will be gathered from them. (3) Two (2) laptops were utilized facing each other with installed prototype data gathering program. (4) The participants were given twenty (20) minutes to answer the programming problem assigned to them. (5) Java Programming Language will be used by the participants in answering the problem. (6) The Prototype Data Gathering Program is already running and it captures the features on the given time. The questionnaire will pop -up every three (3) minutes. (7) The recorded keystrokes and mouse movement and clicks is saved into an excel file. It is then converted to .csv (comma delimited format) to be used as an input operator for data mining.

3.3 Feature Extraction

Feature extraction includes lessening the measure of assets required to portray a vast arrangement of information. When performing analysis in the complex data, one of the real issues comes from the quantity of factors included. So, before starting to train, before beginning to prepare a classifier, it is important to play out an element choice strategy to dispense with the parameters which don't relate with the class names and in this manner, decrease the dimensionality [20]. In this research, the plan is to characterize few profoundly solid components indicating certain feelings angle, for example, pleasant, calm, etc. Every element will identify the emotion level of the user. Name of the participant, age and sex were also extracted together with the keyboard and mouse data. The keystrokes, transition between key press, mouse x position, mouse y position, action, delay (dwelt time per milliseconds) and execution time [19]. The chosen emotional condition of the student is then mapped every time the student uses the mouse and the keyboard and based on the answer from the Self - Assessment Manikin.

From the extracted file, this research has acquired 14, 889 rows of recorded mouse and keyboard data, which will be used for the training and testing using the best algorithm available.

3.4 Model Development

This research developed a model to be used in the prototype program that will identify and analyze the emotion of the user using keyboard and mouse. The extracted feature is then analyzed using free open -source software available in the web like RapidMiner. The researcher selected 80% from the mapped file to be used as the TrainingSet and the remaining 20% is used as the TestSet. The two files was saved as .csv (comma delimited format) to be used as an input data in the RapidMiner. RapidMiner has been chosen by this research because of its user interface that is very friendly. There is more than one approach in labeling emotions and these methods include human labeling and the other is automatic labeling [16]. There are various kinds of labels that people can select from the word labels list from the questionnaire and uses scales to present the level of responses to each question [likert]. The other one is the use of Self-Assessment Manikin (SAM) which arousal and valence are presented and expressed graphically [9, 10]. The SAM is a pictorial evaluation tool intended to know the emotional measurements specifically by methods for two arrangements of graphical image. This method has been used in different studies and application [21, 22].

3.5 Classification

For classification, since the label for the data set is nominal, the researcher used five (5) different classifiers such as Decision Tree, Random Forest, Gradient Boosted Trees, Decision Stump and Random Tree. The result of training is shows that Gradient Boosted Trees presented the most acceptable model which has an accuracy of 70.25% and Kappa of 0.612%.

3.6 Model Testing

Model testing is done to verify that the gradient boosted tree has high accuracy and kappa using the test set data which is the 20% of the gathered data from the participants. The testing shows that the accuracy rate of the test model is 79% which is 13% higher than the Training set with the kappa of 0.75% which is 23% higher than the Training set.

4 Results and Discussion

The machine learning algorithm called Gradient Boosted Trees also known in other terms as XGboost [25], has function called Huber-M Loss generates a series of decision trees for loss data to return fit for classification [21]. The comparison of XGboost to other machine learning algorithms has highest accuracy rate 70.25% and kappa rate 0.61% when used to generate model based on the pattern and classes collected. It is because however, the function Huber-M Loss therein presented allows this algorithm to fit Loss data by generating another trees to the missing values. The accuracy indicates the highest precision point of all the classes. The classified correct example compared to the number of all examples known as accuracy is then calculated as a set of random input $x = \{x1,...xn\}$ using the training set sample $\{Yi, Xi\}$ or known values, and the positives ratio to all as positive predicted examples which is the precision is calculated based on the number of iteration and learning rate. The ratio of positives to all actually is positive examples known as recall is calculated based on splits nonterminal node [21].

Base on XGBoost formula the highest predictions are always on left which means the milliseconds delay value between 0.212 to 0.243 whether Male or Female after more than a second that participant starting, the "PLEASED NEUTRAL" is predicted, and most efficient compare to the right thereby the rest to the right belong to the category (Male/Female) is void.

```
If Gender = Male Then
    If seconds <= 2 Then
        If MouseClicked = LeftClicked Then
            If seconds >= 1 Then
                If delay = 0.212 Then
                    Emotion = "PLEASE NEUTRAL"
                End If
            End If
        End If
    End If
ElseIf Gender = Female Then
    If Mouse = Moved Then
        If delay = 4.5 Then
            If seconds >= 1 Then
                If delay = 0.243 Then
                    Emotion = "PLEASE NEUTRAL"
                End If
            End If
        End If
    End If
End If
```

Fig. 2. Sample prototype

The actual interface of our prototype on the figure below consists of only one button; it designs to be user friendly and straightforward (Fig. 2).

The prototype program automatically generates .XLS and .TXT files when it is started, just by defining its proper directory and filename. The TXT file produce the same data as XLS does but is prepared for retrieving or viewing through web.

It is possible to predict emotion using SAM questionnaire [26], keystrokes and mouse clicks detect stress [18] although this paper uses different subject and procedures but closely similar to the application of keystrokes dynamic [19]. The findings of this research not necessarily good but different where it was tested the students on programming subject, for future works should explore additional parameters such as to check program error or if the student think while doing the programming task how it will affect emotion.

5 Conclusion

Intelligent computers have important roles in society today, with its full capacity like its ability to recognize emotions that can be used to know the needs of the users. Extracting the emotional aspects of the human with the use of computers system could help the decision of the user on how would to interact with the computer. Thus, adapting to this new system that could identify and analyze the emotion of the person using the computer is a big help not only to the academe but also to the general user. The current approaches in emotion identification with the use of computer requires expensive equipment and is intrusive in nature. This paper has presented one way that could be used to identify and analyze the emotion in a non -intrusive and inexpensive way. Using the keyboard and mouse provides an alternative way of detecting the emotions of the students. Identifying these emotions and taking the appropriate actions can be of great help for teachers in assigning computer related activities to students thereby helping in the learning process.

References

1. Gross, J.J., Muñoz, R.F.: Emotion regulation and mental health. Clin. Psychol. Sci. Pract. **2**(2), 151–164 (1995)
2. Salovey, P., Rothman, A.J., Detweiler, J.B., Steward, W.T.: Emotional states and physical health. Am. Psychol. **55**(1), 110 (2000)
3. Picard, R.W.: Affective computing: from laughter to IEEE. IEEE Trans. Affect. Comput. **1**(1), 11–17 (2010)
4. Picard, R.W., Picard, R.: Affective Computing, vol. 252. MIT press, Cambridge (1997)
5. Khanna, P., Sasikumar, M.: Recognizing emotions from keyboard stroke pattern. Int. J. Comput. Appl
6. Khan, I.A., Brinkman, W.P., Hierons, R.: Towards estimating computer users' mood from interaction behaviour with key board and mouse. Front. Comput. Sci. **7**(6), 943–954 (2013)
7. Salmeron-Majadas, S., Santos, O.C., Boticario, J.G.: Exploring indicators from keyboard and mouse interactions to predict the user affective state. Educ. Data Min. (2014)

8. Grimes, G.M., Jenkins, J.L., Valacich, J.S.: Exploring the effect of arousal and valence on mouse interaction. In: 34th International Conference on Information Systems. Milan (2013)
9. Lang, P., Bradley, M.M.: The International Affective Picture System (IAPS) in the Study of Emotion and Attention, Handbook of emotion elicitation and assessment, p. 29 (2007)
10. Lang, P.J.: The emotion probe. Am. Psychol. **50**(5), 372–385 (1995)
11. Bradley, M.M., Greenwald, M.K., Petry, M.C., Lang, P.J.: Remembering pictures: pleasure and arousal in memory. J. Exp. Psychol. Learn. Mem. Cognit. **18**(2), 379–390 (1992)
12. Greenwald, M.K., Bradley, M.M., Hamm, A.O., Lang, P.J.: Looking at pictures: evaluative, facial, visceral and behavioral responses. Psychophysiology **30**(3), 261–273 (1993)
13. Miller, G.A., Levin, D.N., Kozak, M.J., Cook III, E.W., McLean Jr., A., Lang, P.J.: Individual differences in imagery and the psychophysiology of emotion. Cognit. Emot. **1**(4), 367–390 (1987)
14. Salmeron-Majadas, S., Santos, O.C., Boticario, J.G.: An evaluation of mouse and keyboard interaction indicators towards non-intrusive and low cost affective modeling in an educational context. Procedia Comput. Sci. **35**, 691–700 (2014)
15. Ball, G., Breese, J.: Emotion and Personality in a Conversational Character. Embodied Conversational Agents. MIT Press, Cambridge (2000)
16. Bixler, R., D' Mello, S.: Detecting boredom and engagement during writing with keystroke analysis, task appraisals, and stable traits. In: Proceedings of the 2013 International Conference on Intelligent User Interfaces, pp. 225–234. ACM, March 2013
17. Zimmermann, P., Guttormsen, S., Danuser, B., Gomez, P.: Affective computing—a rationale for measuring mood with mouse and keyboard. Int. J. Occup. Safety Ergon. **9**(4), 539–551 (2003)
18. Rodrigues, M., Gonçalves, S., Carneiro, D., Novais, P., Fdez-Riverola, F.: Keystrokes and clicks: measuring stress on e-learning students. In: Casillas, J., Martínez-López, F., Vicari, R., De la Prieta, F. (eds.) Management Intelligent Systems. Advances in Intelligent Systems and Computing, vol. 220, pp. 119–126. Springer, Heidelberg (2013). https://doi.org/10.1007/978-3-319-00569-0_15
19. Shukla, P., Solanki, R.: Web based keystroke dynamics application for identifying emotional state. Institute of Engineering and Technology DAVV, Indore, India, November 2013
20. Kołakowska, A., Landowska, A., Szwoch, M., Szwoch, W., Wróbel, M.R.: Emotion recognition and its application in software engineering. In: 2013 The 6th International Conference on Human System Interaction (HSI), pp. 532–539. IEEE, June 2013
21. Lee, P.M., Tsui, W.H., Hsiao, T.C.: The influence of emotion on keyboard typing: an experimental study using auditory stimuli. PLoS ONE **10**(6), e0129056 (2015)
22. Friedman, J.H.: Greedy function approximation: a gradient boosting machine. Ann. stat. **29**(5), 1189–1232 (2001)
23. Mitra, S., Pal, S.K., Mitra, P.: Data mining in soft computing framework: a survey. IEEE Trans. Neural Netw. **13**(1), 3–14 (2002)
24. Click, C., Malohlava, M., Candel, A., Roark, H., Parmar, V.: Gradient Boosted Models with H2O, March 2016. https://h2o-release.s3.amazonaws.com/h2o/rel-turan/4/docs-website/h2o-docs/booklets/GBM_Vignette.pdf.Amazonaws.com. Accessed M ay 2017
25. Cao, G., Ding, J., Duan, Y., Tu, L., Xu, J., Xu, D.: Classification of tongue images based on doublet and color space dictionary. In: IEEE International Conference on Bioinformatics and Biomedicine (BIBM), pp. 1170–1175 (2016)
26. Mchugh, M.L.: Interrater reliability: the kappa statistic. Biochemia Medica, pp. 276–282 (2012)
27. Bradley, M.M., Lang, P.J.: Measuring emotion: the self-assessment manikin and the semantic differential. J. Behav. Therapy Exp. Psychiatry **25**(1), 49–59 (1994)

28. Smith, T.F., Waterman, M.S.: Identification of common molecular subsequences. J. Mol. Biol. **147**, 195–197 (1981)
29. May, P., Ehrlich, H.C., Steinke, T.: ZIB structure prediction pipeline: composing a complex biological workflow through web services. In: Nagel, W.E., Walter, W.V., Lehner, W. (eds.) Euro-Par 2006. LNCS, vol. 4128, pp. 1148–1158. Springer, Heidelberg (2006)
30. Foster, I., Kesselman, C.: The Grid: Blueprint for a New Computing Infrastructure. Morgan Kaufmann, San Francisco (1999)
31. Czajkowski, K., Fitzgerald, S., Foster, I., Kesselman, C.: Grid information services for distributed resource sharing. In: 10th IEEE International Symposium on High Performance Distributed Computing, pp. 181–184. IEEE Press, New York (2001)
32. Foster, I., Kesselman, C., Nick, J., Tuecke, S.: The physiology of the grid: an open grid services architecture for distributed systems integration. Technical report, Global Grid Forum (2002)
33. National Center for Biotechnology Information. http://www.ncbi.nlm.nih.gov

Waiting Time Screening in Healthcare

José Neves[1]([✉]), Henrique Vicente[1,2] [iD], Marisa Esteves[3] [iD],
Filipa Ferraz[3] [iD], António Abelha[1] [iD], José Machado[1] [iD],
Joana Machado[4] [iD], and João Neves[5] [iD]

[1] Centro Algoritmi, Universidade do Minho, Braga, Portugal
{jneves,abelha,jmac}@di.uminho.pt
[2] Departamento de Química, Escola de Ciências e Tecnologia, Centro de
Química de Évora, Universidade de Évora, Évora, Portugal
hvicente@uevora.pt
[3] Departamento de Informática, Universidade do Minho, Braga, Portugal
marisa.araujo.esteves@gmail.com,
filipatferraz@gmail.com
[4] Farmácia de Lamaçães, Braga, Portugal
joana.mmachado@gmail.com
[5] Mediclinic Arabian Ranches, PO Box 282602, Dubai, United Arab Emirates
joaocpneves@gmail.com

Abstract. In Medical Imaging (MI), various technologies can be used to monitor the human body for diagnosing, monitoring or treating disease. Each type of technology provides different information about the body area that is being investigated or treated for a possible illness, injury or effectiveness of a medical treatment. Routine screening has identified malfunction detection in many otherwise asymptomatic patient images such as computed tomography or magnetic resonance. Studies have shown that, compared to patients whose disease was symptomatic (i.e., self-recognizing), screen-detected diseases may have more favorable clinicopathological features, leading to better prognosis and better outcome. This paper aims to assess the issue of health care wait screening. It deviates from a decision support system that evaluates the waiting times in diagnostic MI based on operational data from various information systems. Last but not least, one's assumptions may have an important impact in determining the usefulness of routine laboratory testing at admission.

Keywords: Waiting time screening · Logic programming
Case-based reasoning

1 Introduction

The characterization of health activities in terms of time-screening theory is a very recent trend in the field of research, i.e., a progressive learning experience, compensated by the occasional satisfaction of discovery. In fact, time-screening has often been defined, with meanings ranging from "not easy" to "persistent." On the other hand, technological advances are rapidly increasing interoperability, i.e., the ability to communicate and integrate information from heterogeneous sources or services. In fact,

© ICST Institute for Computer Sciences, Social Informatics and Telecommunications Engineering 2018
J. J. Jung et al. (Eds.): BDTA 2017, LNICST 248, pp. 124–131, 2018.
https://doi.org/10.1007/978-3-319-98752-1_14

a variety of imaging techniques can be used to diagnose or treat diseases such as X-rays, Computed Tomography (CT), Magnetic Resonance Imaging, Positron Emission Tomography, Nuclear Medicine. Thus, a large data set extracted from various information systems such as the Radiology Information System, the Image Archiving and Communication System and the Electronic Medical Record is acquired and processed [1–3]. It is undeniable that a proactive strategy is needed to solve such a problem delay, which must take all of these factors into account. It is from this point of view that the problem is addressed. The basis of integrated care is to be understood as a patient interacting with a prepared, proactive and multidisciplinary setting. In this work, the approach focuses on estimating the waiting time in a Case Based Reasoning (CBR) approach to problem solving [4–6].

2 Knowledge Representation and Reasoning (KRR)

One aims at the understanding of the information's complexity and the associated inference mechanisms. Indeed, automated reasoning capabilities enables a system to fill in the blanks when one is dealing with incomplete information, where data gaps are common. In this study, a data item is to be understood as find something smaller inside when taking anything apart, i.e., it is mostly formed from different elements, namely the *Interval Ends* where their values may be situated, the *Quality-of-Information* (*QoI*) they carry, and the *Degree-of-Confidence* (*DoC*) put on the fact that their values are inside the intervals just referred to above. These are just three of over an endless element's number. Undeniably, one can make virtually anything one may think of by joining different elements together or, in other words, viz.

- What happens when one splits a data item? The broken **pieces** become data item for another element, a process that may be endless; and
- Can a data item be broken down? Basically, it is the smallest possible part of an element that still remains the element.

Therefore, the proposed approach to this issue, put in terms of the logical programs that elicit the universe of discourse, will be set as productions of the type, viz

$$predicate_{1 \leq i \leq n} - \bigcap_{1 \leq j \leq m} clause_j(([A_{x_1}, B_{x_1}](QoI_{x_1}, DoC_{x_1})), \cdots$$

$$\cdots, ([A_{x_m}, B_{x_m}](QoI_{x_m}, DoC_{x_m}))) :: QoI_j :: DoC_j$$

where n, $\cap$, m and A_{x_m}, B_{x_m} stand for the cardinality of the predicates' set, conjunction, predicate's extension, and the interval ends where the predicates attributes values may be situated, respectively. The metrics $[A_{x_m}, B_{x_m}]$, QoI and DoC show the way to data item dissection, i.e., a data item is to be understood as the data's atomic structure. It consists of identifying not only all the sub items that are thought to make up an data item, but also to investigate the rules that oversee them, i.e., how $[A_{x_m}, B_{x_m}]$, QoI_{x_m}, and DoC_{x_m} are kept together and how much added value is created [7–14].

3 Case Study

A database was set to create an intelligent system for the planning process of *Waiting Time Screening in Healthcare*. The knowledge database is composed by a set of predicate's extensions (Fig. 1). Some incomplete or default data are present under this scenario (for instance, the *type* in case 1 is unknown, and symbolized as $\perp$).

					Waiting Time Screening				
Attributes of the Feature Vector:	#	Age	Gender	Date (days)	Modality	Type	Priority	Ordering Speciality	Description
Feature Vector Attributes' Values:	1	37	0	17	1	$\perp$	0	27	Description 1
	2	77	1	45	1	55	1	92	Description 2
	...	...	...	...	...	...	...	...	...
	149	68	0	21	0	83	0	43	Description 149
Feature Vector Domains:		[17, 95]	[0, 1]	[1, 366]	[0, 1]	[0, 113]	[0, 1]	[0, 192]	

Fig. 1. Healthcare table or relation.

The table has several columns such as *Gender, Modality* and *Priority* of *Waiting Time Screening*. The table rows have been inserted with one (1) or 0 (zero) standing for, respectively, *Male/Female, CT/MRI* and *Urgent/Routine*.

It is now possible to define the predicate *waiting time screening* (*wts*) whose extension stands for the objective function with respect to the problem under analysis, viz.

$$wts : Age, G_{ender}, Date, Mod_{ality}, T_{ype}, Prio_{rity}, O_{rdering}S_{pecialty} \rightarrow \{0, 1\}$$

in which the truth values *true* and *false* are expressed by 1 (one) and 0 (zero), respectively. Considering the feature vector ($Age = 44$, $G_{ender} = 1$, $Date = [30, 45]$, $Mod_{ality} = 0$, $T_{ype} = 69$, $Prio_{rity} = 1$, $O_{rdering}\ S_{pecialty} = \perp$), one may have, viz.

$$\{$$

$$\neg wts\left(\left((A_{Age}, B_{Age})(QoI_{Age}, DoC_{Age})\right), \cdots, ((A_{OS}, B_{OS})(QoI_{OS}, DoC_{OS}))\right)$$

$$\leftarrow wts\left(\left((A_{Age}, B_{Age})(QoI_{Age}, DoC_{Age})\right), \cdots, ((A_{OS}, B_{OS})(QoI_{OS}, DoC_{OS}))\right)$$

$$wts\underbrace{\left(((0.35, 0.35)(1, 1)), \cdots, ((0, 1)(1, 0))\right)}_{\substack{\text{attribute's values ranges once normalized and}\\ \text{respective QoI and DoC values}}} :: 1 :: 0.86$$

$$\underbrace{[0, 1] \qquad \cdots \qquad [0, 1]}_{\substack{\text{attribute's domains}\\ \text{once normalized}}}$$

$$\} :: 1$$

Program 1. The *Logic Program* with respect to the feature vector referred to above.

4   Case Based Reasoning (CBR)

The *CBR* cycle used in this work was proposed by Neves *et al.*, [12, 15] (Fig. 2), with the ability to deal with incomplete or unknown information. Artificial Neural Networks (ANNs) [16] were used in the optimization stage. The value ranges boundaries of the attribute, the *DoCs* and *QoIs*, are the inputs of the *ANN*. The output not only provides the case assessment, but also a confidence measure that deals with such a categorization (Fig. 3).

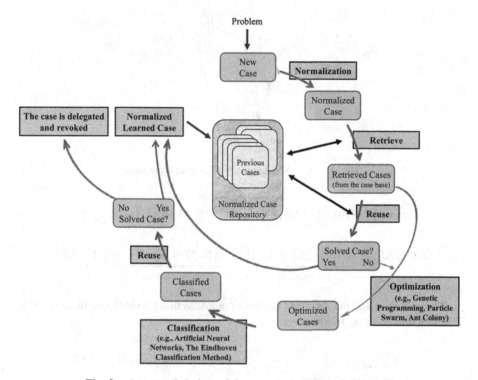

Fig. 2.   An extended view of the canonical *CBR* cycle [12, 15].

When faced with a *new case*, for example the one that presents feature vector $Age = 56$, $Gender = 0$, $Date = 21$, $Mod_{ality} = \perp$, $Type = 42$, $Prio_{rity} = 0$, $O_{rdering}$ $S_{pecialty} = 112$, one may have, viz.

$$wts_{newcase} \underbrace{(((0.5, 0.5)(1, 1)), \cdots, ((0.58, 0.58)(1, 1)))}_{\substack{attribute's \ values \ ranges \ once \ normalized \\ and \ respective \ QoI \ and \ DoC \ values}} :: 1 :: 0.85$$

leading to a retrieving of 42 cases [17], viz.

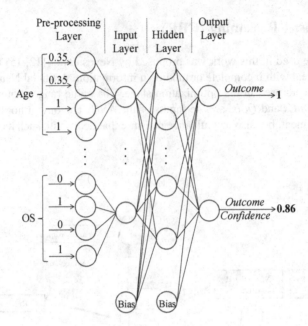

Fig. 3. The *ANNs* approach to case optimization.

$$retrieved_{case_1}((((0.43, 0.43)(1, 1)), \cdots, ((0.86, 0.86)(1, 1)))) :: 1 :: 0.83$$

$$\vdots$$

$$retrieved_{case_{42}} \underbrace{(((0.51, 0.51)(1, 1)), \cdots, ((0.57, 0.57)(1, 1)))}_{normalized\ cases\ that\ make\ the\ retrieved\ cluster} :: 1 :: 0.84$$

The *new case* and the *retrieved ones* are compared using a similarity function, *sim*. This function is set as follows, viz.

$$sim^{DoC}_{newcase \to 1} = 1 - \frac{\|1 - 1\| + \cdots + \|1 - 1\|}{7} = 1 - 0.18 = 0.82$$

where $sim^{DoC}_{newcase \to 1}$ stands for the *similarity* with respect to *DoC*, between the *new case* and the retrieved ones (in this example *retrieved case_1*). A similar process was considered in order to evaluate the *similarity*, in terms of *QoI*, between the *new case* and the *retrieved case_1*, returning $sim^{QoI}_{newcase \to 1} = 1$. The *general similarity*, $sim^{QoI,DoC}_{newcase \to 1}$, is the product of the above metrics above, viz.

$$sim^{QoI,DoC}_{newcase \to 1} = 1 \times 0.82 = 0.82$$

This method was extended to all the remaining cases leading to the most similar case, i.e., the potential problem solutions. The coincidence matrix for the *CBR* model is shown in Table 1. It shows that the *CBR* model classifies properly 133 of a total of 149 cases, being the model accuracy 89.2%. In terms of the well known statistical metrics

Table 1. The coincidence matrix

Output	Model output	
	True (1)	False (0)
True (1)	92	10
False (0)	6	41

such as sensitivity and specificity, the results were 90.2%, and 87.2% respectively. The *ROC* curve is shown in Fig. 4. The area under the curve is 0.89. The performance metrics [18, 19] are close to 90% and suggest that the model has a good performance in predicting the waiting time in healthcare.

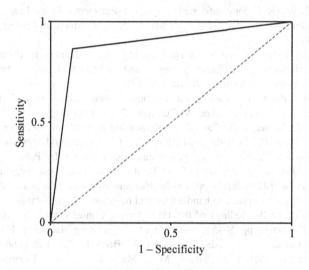

Fig. 4. The *ROC* curve.

5   Conclusions

This work begins with the development of an intelligent system for assessing the latency in providing diagnostic medical services, based on a formal framework based on Logic Programming for knowledge representation and reclaiming the CBR approach to problem solving. The knowledge presentation and enforcement apparatus presented above is very versatile and able to cover all possible data types, namely incomplete, unknown or even self-contradictory data or information. Future work should include data from various healthcare facilities (public, semi-public and private) from different regions of Portugal. On the other hand, different string similarity strategies will be considered, and their analysis complexity enumerated. On the other hand, given the cost of this relatively low score, these results have important implications for the doctor's office and cost-benefit analyzes that will be further evaluated to better determine the current benefit of routine laboratory testing on admission.

Acknowledgments. This work has been supported by COMPETE: POCI-01-0145-FEDER-007043 and FCT – Fundação para a Ciência e Tecnologia within the Project Scope: UID/CEC/00319/2013.

References

1. Nuti, S., Vainieri, M.: Managing waiting times in diagnostic medical imaging. BMJ Open **2**, e001255 (2012)
2. McEnery, K.W.: Radiology information systems and electronic medical records. In: IT Reference Guide for the Practicing Radiologist, pp. 1–14. American College of Radiology, USA (2013)
3. Fotiadou, A.: Choosing and visualizing waiting time indicators in diagnostic medical imaging department for different purposes and audiences. Master's thesis in Health Informatics, Karolinska Institutet, Sweden (2013)
4. Aamodt, A., Plaza, E.: Case-based reasoning: foundational issues, methodological variations, and system approaches. AI Commun. **7**, 39–59 (1994)
5. Richter, M.M., Weber, R.O.: Case-Based Reasoning: A Textbook. Springer, Berlin (2013)
6. Balke, T., Novais, P., Andrade, F., Eymann, T.: From real-world regulations to concrete norms for software agents – a case-based reasoning approach. In: Poblet, M., Schild, U., Zeleznikow, J. (eds.) Proceedings of the Workshop on Legal and Negotiation Decision Support Systems (LDSS 2009), pp. 13–28. Huygens Editorial, Barcelona (2009)
7. Neves, J.: A logic interpreter to handle time and negation in logic databases. In: Muller, R., Pottmyer, J. (eds.) Proceedings of the 1984 Annual Conference of the ACM on the 5th Generation Challenge, pp. 50–54. Association for Computing Machinery, New York (1984)
8. Neves, J., Machado, J., Analide, C., Abelha, A., Brito, L.: The halt condition in genetic programming. In: Neves, J., Santos, M.F., Machado, J. (eds.) Progress in Artificial Intelligence. LNAI, vol. 4874, pp. 160–169. Springer, Berlin (2007)
9. Kakas, A., Kowalski, R., Toni, F.: The role of abduction in logic programming. In: Gabbay, D., Hogger, C., Robinson, I. (eds.) Handbook of Logic in Artificial Intelligence and Logic Programming, vol. 5, pp. 235–324. Oxford University Press, Oxford (1998)
10. Pereira, L., Anh, H.: Evolution prospection. In: Nakamatsu, K. (ed.) New Advances in Intelligent Decision Technologies, Studies in Computational Intelligence, vol. 199, pp. 51–64. Springer, Berlin (2009)
11. Machado, J., Abelha, A., Novais, P., Neves, J., Neves, J.: Quality of service in healthcare units. In Bertelle, C., Ayesh, A. (eds.) Proceedings of the ESM 2008, pp. 291–298. Eurosis – ETI Publication, Ghent (2008)
12. Silva, A., Vicente, H., Abelha, A., Santos, M.F., Machado, J., Neves, J., Neves, J.: Length of stay in intensive care units – a case base evaluation. In: Fujita, H., Papadopoulos, G.A. (eds.) New Trends in Software Methodologies, Tools and Techniques, Frontiers in Artificial Intelligence and Applications, vol. 286, pp. 191–202. IOS Press, Amsterdam (2016)
13. Fernandes, F., Vicente, H., Abelha, A., Machado, J., Novais, P., Neves, J.: Artificial neural networks in diabetes control. In: Proceedings of the 2015 Science and Information Conference (SAI 2015), pp. 362–370, IEEE Edition (2015)

14. Turner, M., Fauconnier, G.: Conceptual integration and formal expression. J. Metaphor Symbolic Act. **10**, 183–204 (1995)
15. Vilhena, J., Vicente, H., Martins, M.R., Grañeda, J., Caldeira, F., Gusmão, R., Neves, J., Neves, J.: A case-based reasoning view of thrombophilia risk. J. Biomed. Inf. **62**, 265–275 (2016)
16. Haykin, S.: Neural Networks and Learning Machines. Pearson Education, New Jersey (2009)
17. Figueiredo, M., Esteves, L., Neves, J., Vicente, H.: A data mining approach to study the impact of the methodology followed in chemistry lab classes on the weight attributed by the students to the lab work on learning and motivation. Chem. Educ. Res. Pract. **17**, 156–171 (2016)
18. Florkowski, C.: Sensitivity, specificity, receiver-operating characteristic (ROC) curves and likelihood ratios: communicating the performance of diagnostic tests. Clin. Biochem. Rev. **29**(Suppl 1), S83–S87 (2008)
19. Hajian-Tilaki, K.: Receiver operating characteristic (ROC) curve analysis for medical diagnostic test evaluation. Caspian J. Intern. Med. **4**, 627–635 (2013)

Analysis and Visualization of University Twitter Feeds Sentiment

Arlene Caballero[1(✉)], Jasmin D. Niguidula[2],
and Jonathan M. Caballero[2]

[1] Lyceum of the Philippines University Intramuros Manila, Manila, Philippines
arlene.caballero@lpu.edu.ph
[2] Technological Institute of the Philippines, 1328 Arlegui Street, Quiapo,
Manila, Philippines
jasniguidula@yahoo.com, jonathanmcaballero@gmail.com

Abstract. The exponential growth of online social network as communication channel brought revolutionary changes in our daily lives. For the organizations, Twitter can be used for many reasons. It can be used as a channel of communication for expressing thoughts, emotions, experiences, perspectives, and opinions in a variety of topics and social interest. This study focused on the information theoretic review of sentiment analysis and visualization in Twitter. This paper examined the tweeter feeds from a select institution using a web-based sentiment analysis tool for analyzing tweeter sentiment and tweet visualization. This further investigates the clustering techniques and information theory applied to visualize and analyze the sentiments in the tweeter feeds using a query as the target of sentiments performed on over 1,500 tweeter feeds from a select institution users. In this study, the individual tweets from the users were converted into images and presented in forms of charts, graphs and diagrams to discover the nature of activity of the users. In view of this, an approach to data mining technique – Shannon information theory has been examined to analyze and review how the estimated sentiment in the corpus of data extracted from the tweeter feeds were processed and calculated. The tf-idf calculated for each query term in tweeter feeds were converted into images using information theoretic approach. With this, the nature of activity and opinions of the users in a select institution were discovered. This study also described the tweeter sentiments in an emotional scatter diagram mapped with pleasure and stimulation using the Russel Model of Affect.

Keywords: Social networking · Opinion mining · Twitter profile
Information theory · Term frequency-inverse document frequency

1 Introduction

Twitter is an online social networks which became one of the most popular micro-blogging services available in the web. As of midyear 2011, over 200 million tweets has been posted in Twitter per day and this has been the subject of attention of researchers of various organizations worldwide [1, 2]. The exponential growth of online social network as communication channel brought revolutionary changes in our

© ICST Institute for Computer Sciences, Social Informatics and Telecommunications Engineering 2018
J. J. Jung et al. (Eds.): BDTA 2017, LNICST 248, pp. 132–145, 2018.
https://doi.org/10.1007/978-3-319-98752-1_15

daily lives [3]. For the organizations, Twitter can be used for many reasons. It can be used as a channel of communication for expressing thoughts, emotions, experiences, perspectives, and opinions in a variety of topics and social interest [3, 4]. Consequently, it serves as a valuable source of public opinion and communication [4].

Opinion Mining also called sentiment analysis, is the field of study that analyses people's opinions, sentiments, evaluations, appraisals, attitudes, and emotions towards entities such as products, services, organizations, individuals, issues, events, topics, and their attributes [5]. It aims to pay attention and process the data being posted by the users in a social media [2].

Sentiment Visualization converts twitter sentiments into images which allow the viewer to see the values and the relationships of data as they form [6]. In this study, the data from a select institution represented by individual tweets from the users were converted into images and presented in forms of charts, graphs and diagrams to discover the nature of activity of the users. In view of this, an approach to data mining technique – Shannon information theory has been examined to analyze and review how the estimated sentiment in the corpus of data extracted from the tweeter feeds were processed and calculated.

This paper utilized and described the tweeter feeds from a select institution using *SentimentViz* - a web-based sentiment analysis tool for analyzing tweeter sentiment and tweet visualization [7]. This further investigates the clustering techniques and information theory applied to visualize and analyze the sentiments in the tweeter feeds using a query as the target of sentiments performed on over 1,500 tweeter feeds from a select institution users.

2 Related Works

This section reviews Shannon information theory and how it relates and intersects to other field of studies such as mathematical or probability theory and algorithmic complexity. This section relates the computational theory applied in sentiment visualization and calculating the estimated sentiment of the twitter feeds.

2.1 Shannon Information Theory

In the early development of human communication, people used pictures, scripts, codes, and symbols to represent objects and to communicate their ideas, actions, names, or by association. In Shannon "theory of communication" which in the latter called "theory of information" explains that the transmission of symbols, script, or message involves sending of information through electronic signals. This communication model consists of transmitter, channel, and a receiver. In the information source, message is being transmitted in a form of signal then, the received signal is sent to the destination [8].

This is further refined in the standard communication model as the source or encoder which translates the message into codes in the form of bits. A code is a set of symbols or a language that is used to transmit message or thought on one or more channel to get response in a receiver or decoder [8].

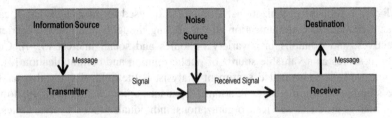

Fig. 1. Shannon model of information theory [8]

Figure 1 show how the message is being transmitted to the receiver through a channel. Whenever a message is transmitted, the noise source which refers to a number of variables makes the message to be distorted or changed. Hence, makes the recipient appear in a different state. These changes refers to entropy which measures the amount of uncertainty of an unknown or random quantity [8]. This communication problem is exactly what Shannon entropy theory tries to measure using the formulation:

$$H(X) = -\sum p(x) \, log_2 \, p(x)$$

Over the years, Shannon theory has grown and evolved into modern communication from the attempts of several mathematicians and communication engineers to define and establish how information source is being communicated and how it can be measured [9]. In the same context, messages can also be transmitted through a form of social media such as Twitter and then convey these messages in a form of data visualization through digital and numeric calculation.

Figure 2 show that relationship of information theory to other fields [10]. In the area of computer science and probability theory, Shannon's quantification of entropy initiate the ideas for a relationship with term frequency-inverse document frequency (tf-idf) applied in this study to weight terms in the corpus of data extracted from the tweeter feeds.

2.2   Term Frequency-Inverse Document Frequency (tf-idf)

One of the most commonly used measures in information retrieval is the "term frequency-inverse document frequency (td-idf). This information weighing schemes is used to measure the probability-weighted amount of information in a given document [11].

In the conventional information theory, idf can be interpreted as 'the amount of information' given as the log of the inverse probability [11, 12]. By definition, tf-idf is a measure that multiples the two quantities tf and idf. With this, term frequency provides estimation of the occurrences probability of a term when it is normalized by the total frequency in the document, or the document collection, depending on the scope of the calculation.

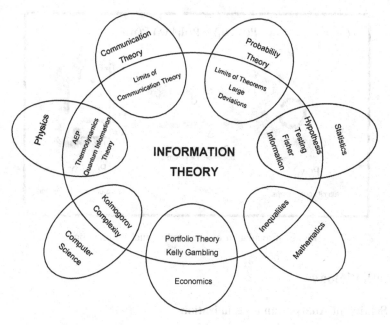

Fig. 2. Relationship of information theory to other fields

Based on the basic formula of information theory [8, 10], a document is assumed to be a given unordered set of terms. Let $D = \{d_j, ..., d_n\}$ be a set of documents and $W = \{w_i, ..., w_M\}$ be as set of distinct terms contained in D. In this study, the documents D is represented by a corpus of data extracted from the tweeter feeds while W is the query term. The parameters N are the total numbers of documents while M are the number of terms. In adapting the theory, selection of term w_i from W and selection of document d_j from D are also considered [11].

To illustrate the probability distribution of terms in a set of documents, the amount of information is calculated as illustrated in the figure below.

Figure 3 illustrates the calculation of expected mutual information when the user submits a query term to process and how the document is being selected using the term. In the estimate, the probability of query terms is represented by $P(w_i)$ while the probability distribution on the selection of documents is represented by $P(d_j \mid w_i)$. In the process, the co-occurrences of documents and terms are calculated [11].

In this study, the text clustering of Twitter feeds were classified using the theory of term frequency - inverse document frequency (TF-IDF). This measure is applied to evaluate the importance of word to a document in a given data set. The number of times a word appear in each tweet is proportional in the increase of importance and is offset by the frequency of word in a data set [7]. The goal of tf-idf technique for text analysis is to organize a document using a query term. This technique makes each tweet to be classified into sentiments explored in this study.

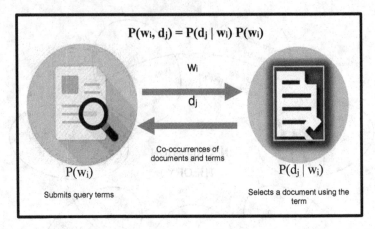

Fig. 3. Calculation of expected mutual information

3   Methodology

3.1   Sentiment Analysis and Visualization

Sentiment analysis is extracting and understanding human emotions from any given text, message or data. In general, sentiment analysis aims to classify documents into polarity of tweets such as positive, negative, or neutral. However, the tf-idf classification technique used in this study aims to classify documents into categories based on query terms. Then, map the terms by emotions using the Russel Model of Affect. The model used in this study proposed the use of valence (or pleasure) and arousal (or stimulation) represented in 2 dimensional plane to build emotional interpersonal circle of affect [7].

Figure 4 depicts the Russel Model of Affect which uses 2D plane emotional dimensions to position feelings or emotions. Along the vertical axis represents arousal (or stimulation) with intense or active on the upper quadrants and mild or passive on the lower quadrants with different levels of pleasures in between. On the horizontal axis, highly unpleasant and highly pleasant were mapped with different levels of pleasure in between. This model suggested using *valence* (or pleasure) and *arousal* (or stimulation) to build emotional interpersonal circle of affect [7].

3.2   Text Clustering Classification

For text clustering, the term frequency - inverse document frequency (tf-idf) was applied to weight the importance of word in a document. Initially, the normalized term frequency (tf) is computed. This refers to the number of times a query term appears in a given data set. Then, the inverse document frequency (idf) is computed which refers to

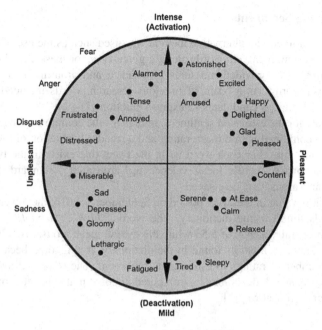

Fig. 4. The Russel Model of Affect

the logarithm of the number of documents where the query term appears. The following are the formula used to calculate tf-idf:

1. *tf (t) = (number of times a query term appears in a document) / (total number of query terms in the document).*

2. *idf (t) = log_e(total number of documents / number of documents with query term in it).*

3. *tf-idf = (term frequency * inverse document frequency)*

In computing the tf, all query terms from the select institution tweet feeds are assumed to be equally important. However, there are terms such as "the", "is", and "of" which may appear often but not as important as the query term. Therefore, an inverse document term frequency is computed to balance down the frequent terms and measure the non-frequent terms. Then, the tf-idf is computed by multiplying the result of term frequency (tf) and the inverse document frequency (idf).

The query term in each tweet feed calculated to have the highest tf-idf result will be selected for the estimate of sentiment. After weighing each term, the standard cosine similarity is applied to compute the pairwise similarity between all pairs of tweets from the corpus of data. The cosine similarity is used to compute for the Dot Product of two vectors which refers to the position of one point in a plot relative to another.

3.3   Estimating Sentiments

In estimating sentiment, an alternative method was used such as the use of dictionaries to determine the sentiments on words along a given tweet or message. The dictionary used *in SentimentViz* provided measures of valence and arousal for approximately 10,680 English words. As a result of previous research, words included in the dictionary were selected as candidates to express emotion [7].

To compute for the estimated sentiment, ratings for the common words found in the dictionary are combined with a mean rating and a standard deviation of the ratings for each dimension rating. For each word wi in the tweet that exists in the tweeter visualizer tool dictionary, word's mean valence $\mu_{v,i}$ arousal $\mu_{a,i}$ standard deviation of valence $\sigma_{v,i}$ and arousal $\sigma_{a,i}$ are saved.

Table 1 shows that in the tweet posted on September 1, 2016, there were more than two (2) words found in the dictionary. The words *"high"*, *"school"*, *"open"*, and *"low"* has an overall valence of 5.55 while the overall mean arousal is 5.30. If a tweet contains less than $n = 2$ words found in the dictionary, it is ignored because it has an insufficient number of ratings to estimate its sentiment. The statistical average of the n means and standard deviations is computed to obtain the tweet's overall mean valence M_v and arousal M_a [7].

Table 1. Tweet valence and arousal

Date/Time Sep 1, 2016 4:01am	'Enrollment for LPU Manila Senior *high School* is NOW *OPEN!* For a *low* downpayment of P5;000'				
Keywords	Mean Valence	Standard Deviation	Mean Arousal	Standard Deviation	Frequency
high	μ: 6.64	σ: 1.21	μ: 4.75	2.91	fq = 50
school	μ: 6.26	σ: 1.88	μ: 5.74	σ: 2.46	fq = 50
open	μ: 6.1	σ: 1.36	μ: 5.92	σ: 2.55	fq = 50
low	μ: 3.66	σ: 1.12	μ: 4.54	σ: 3.19	fq = 50
Valence			v = 5.55		
Arousal			a = 5.30		

3.4   Sentiment Visualization

In visualization of sentiments, each tweet's estimated sentiment is represented by a circle mapped by emotions. An unpleasant tweet is plotted in blue circles while pleasant tweets are mapped in green circles. The stimulation or arousal is represented as brighter circles which indicate that the brighter the circle, the more active are the tweets. The confidence in the sentiment estimate is represented by the size and transparency. The larger the sizes of the circle, the more confident are the estimates.

Another measure of confidence of the tweet's emotion is the transparency. The more opaque or less transparent tweets, the more confident are the estimates.

Figure 5 illustrates how a single tweet with overall mean valence of *5.55* and overall mean arousal of *5.30* is being plotted in an emotional map in horizontal and vertical axes of pleasure and arousal. It further shows that the tweet lies on the upper right quadrant in the Russel Model which depicts that the tweet is generally pleasant.

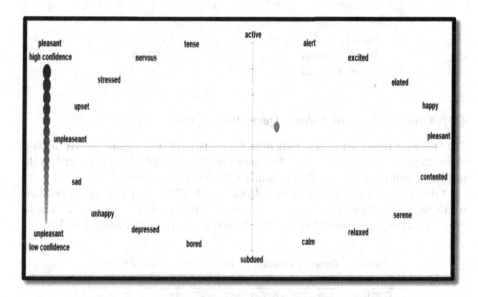

Fig. 5. Tweeter visualization

4   Results and Discussions

4.1   Select Institution Profile

This section describes the profile of the select institution in terms of frequency of posts and number of user interactions. The corpus of data examined in this study were gathered from the tweet feeds of a select institution from December 1, 2011 to June 29, 2016. The tweets were first analyzed by the average frequency per day, user mentions, and hashtags; percentage of retweets, and replies; and percentage of tweets retweeted, and retweets favorited.

As shown in Table 2, the analysis was performed over 1,557 tweets as corpus of data subject for estimated sentiment. On the average, there was 0.93 mean frequency of tweets posted each day and about 655 user mentions with 0.42 average number of mentions per tweet, 1087 hashtags with 0.70 average number of hashtags, 83 retweets or 5% retweets and 361 replies or 23% replies in the total of analyzed tweets. Tweets retweeted has a frequency of 967 or 62.1% while tweets favorited has frequency of 1174 or 75.4%. This shows that the users has more interaction with each other as the numbers calculated increases.

Table 2. Tweets analytics of the select institution

Total of 1557 Tweets
From December 1, 2011 to June 29, 2016

Tweet activity	Frequency	Mean average	Percentage
Tweets per day	0.93		
User Mentions	655	0.42	
Hashtags	1087	0.70	
Retweets	83		5%
Replies	361		23%
Tweets retweeted	967		62.1%
Tweets favorited	1174		75.4%

4.2 Tweets Calculation Using a Query 'Enrollment'

The following tables were the results from the query term 'Enrollment' or 'Enrolment" as the target of sentiments. The table shows the highlighted keywords found in the dictionary and the corresponding measures of valence and arousal.

Table 3 depicts that the tweet feed dated *June 8, 2016, 9:51 pm* highlighted keywords such as *"situation", "best", "option", "enroll", "will", "be", and "enrollment"* which has an overall valence of 6.24 and an overall mean arousal of 3.86.

Table 3. Tweets on enrollment dated June 8, 2016, 9:51 pm

Date/Time Jun 8, 2016 9:51pm	@lxmntrsh_ if they are having *situation*; the *best option* is to *enroll* in LPU and they *will be* assisted by *enrollment* advisers				
keywords	Mean Valence	Standard Deviation	Mean Arousal	Standard Deviation	Frequency
situation	μ: 5.0	σ: 1.31	μ: 4.08	σ: 1.92	fq = 50
best	μ: 7.18	σ: 1.69	μ: 4.6	σ: 2.67	fq = 50
option	μ: 6.49	σ: 1.31	μ: 4.74	σ: 2.23	fq = 5
enrol	μ: 6.19	σ: 1.94	μ: 3.76	σ: 2.63	fq = 23
will	μ: 6.83	σ: 2.04	μ: 2.76	σ: 2.05	fq = 19
be	μ: 6.18	σ: 1.44	μ: 3.43	σ: 2.31	fq = 21
enrollment	μ: 6.19	σ: 1.94	μ: 3.76	σ: 2.63	fq = 23
Valence			v = 6.24		
Arousal			a = 3.86		

Table 4 shows that the tweet feed dated *Oct 11, 2016 1:44am* highlighted keywords such as *"Enroll", "students", "deficiencies", "Please", "See", and "notice"* which has an overall valence of 5.35 and an overall mean arousal of 4.79.

Table 4. Tweets on enrollment dated Oct 11, 2016 1:44 am

Date/Time Oct 11, 2016 1:44am	#*Enrolment* Schedule is now up at http://t.co/zcX5MI0ZCF; For *students* w/ *deficiencies*; *please see notice* there also.@LPUPirates @TagaLPU				
keywords	Mean Valence	Standard Deviation	Mean Arousal	Standard Deviation	Frequency
Enroll	μ: 6.19	σ: 1.94	μ: 3.76	σ: 2.63	*fq* = 23
students	μ: 6.28	σ: 1.83	μ: 5.12	σ: 2.46	*fq* = 50
deficiencies	μ: 2.74	σ: 1.24	μ: 4.2	σ: 2.53	*fq* = 19
please	μ: 6.36	σ: 1.68	μ: 5.44	σ: 2.88	*fq* = 50
see	μ: 6.06	σ: 1.06	μ: 6.1	σ: 2.19	*fq* = 50
notice	μ: 5.16	σ: 1.5	μ: 3.93	σ: 2.56	*fq* = 50
	Valence		*v* = 5.35		
	Arousal		*a* = 4.79		

Table 5. Tweets on enrollment dated Oct 16, 2016 1:26 am

Date/Time Oct 16, 2016 1:26am	Do you have any concern about # Enrollment? Just send a PM to http://t.co/T9cG1fdlJJ and we'd be glad to address them. #OnlineHelpDesk				
keywords	Mean Valence	Standard Deviation	Mean Arousal	Standard Deviation	Frequency
concern	μ: 4.04	σ: 1.62	μ: 5.07	σ: 2.74	*fq* = 50
Enrollment	μ: 6.19	σ: 1.94	μ: 3.76	σ: 2.63	*fq* = 23
send	μ: 5.38	σ: 1.35	μ: 5.63	σ: 2.36	*fq* = 50
be	μ: 6.18	σ: 1.44	μ: 3.43	σ: 2.31	*fq* = 21
glad	μ: 7.48	σ: 1.52	μ: 6.49	σ: 2.77	*fq* = 50
address	μ: 5.6	σ: 1.05	μ: 5.62	σ: 2.25	*fq* = 50
	Valence		*v* = 5.80		
	Arousal		*a* = 4.98		

Table 5 shows that the tweet feed dated *Oct 16, 2016 1:26am* highlighted keywords such as *"concern"*, *"enrollment"*, *"Send"*, *"Be"*, *"glad"*, *and "address"* which has an overall valence of 5.80 and an overall mean arousal of 4.98.

Table 6 depicts that the tweet feed dated *Oct 16, 2016 9:23am* highlighted keywords such as *"thank"*, *"will"*, *"Be"*, *"address"*, *"concerns"*, *and "enrollment"* which has an overall valence of 5.84 and an overall mean arousal of 4.12.

Table 7 depicts that the tweet feed dated *Oct 17, 2016 1:29am* highlighted keywords such as *"please"*, *"check"*, *"website"*, *"enrollment"*, *"students"*, *and "follow"* which has an overall valence of 6.19 and an overall mean arousal of 4.67.

4.3 Tweeter Sentiment Visualization

The tweet feeds as a result of the query term *"Enrollment"* were estimated and mapped in a scatterplot diagram to visualize the sentiments.

Table 6. Tweets on enrollment dated Oct 16, 2016 9:23 am

Date/Time Oct 16, 2016 9:23am	#OnlineHelpDesk *Thank* you for your queries; we *will be* around to *address* your *concerns* throughout *enrollment*; Monday to Friday; 9am - 5pm				
keywords	Mean Valence	Standard Deviation	Mean Arousal	Standard Deviation	Frequency
thank	μ: 6.89	σ: 2.29	μ: 4.34	σ: 2.31	fq = 6
will	μ: 6.83	σ: 2.04	μ: 2.76	σ: 2.05	fq = 19
Be	μ: 6.18	σ: 1.44	μ: 3.43	σ: 2.31	fq = 21
address	μ: 5.6	σ: 1.05	μ: 5.62	σ: 2.25	fq = 50
concerns	μ: 4.06	σ: 1.56	μ: 5.07	σ: 2.74	fq = 50
enrollment	μ: 6.19	σ: 1.94	μ: 3.76	σ: 2.63	fq = 23
Valence			v = 5.84		
Arousal			a = 4.12		

Table 7. Tweets on enrollment dated Oct 17, 2016 1:29 am

Date/Time Oct 17, 2016 1:29am	*Please check* LPU *Website* (http://t.co/51zfnHyX) for the schedule of 2ns Semester *enrollment*. *Students* must *follow* the schedules accordingly.				
keywords	Mean Valence	Standard Deviation	Mean Arousal	Standard Deviation	Frequency
please	μ: 6.36	σ: 1.68	μ: 5.44	σ: 2.88	fq = 50
check	μ: 6.1	σ: 1.53	μ: 6.1	σ: 2.19	fq = 50
website	μ: 6.76	σ: 1.64	μ: 3.58	σ: 2.22	fq = 22
enrollment	μ: 6.19	σ: 1.94	μ: 3.76	σ: 2.63	fq = 23
students	μ: 6.28	σ: 1.83	μ: 5.12	σ: 2.46	fq = 50
follow	μ: 5.66	σ: 1.17	μ: 4.1	σ: 2.12	fq = 50
Valence			v = 6.19		
Arousal			a = 4.67		

Figure 6 demonstrates the visualization of the tweeter feeds streamed from the corpus of data using the query term *"Enrollment"* as a target of sentiments. It further revealed that the tweets lies on the right side of the scatterplot diagram which indicates that the tweets using the query term *"Enrollment"* are generally pleasant.

Figure 7 demonstrates the general tweeter sentiments of the select institution visualized in an emotional scatter diagram. It depicts that the general estimated sentiment reclined on the positive emotions where majority of the sentiments represented by circles in color green are pleasant [7].

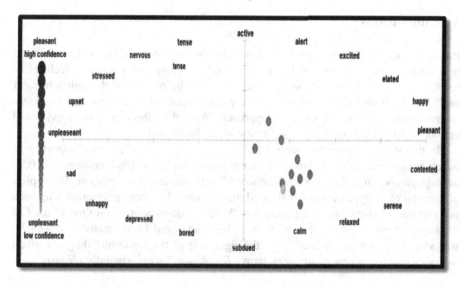

Fig. 6. Tweeter sentiments on query term "Enrollment"

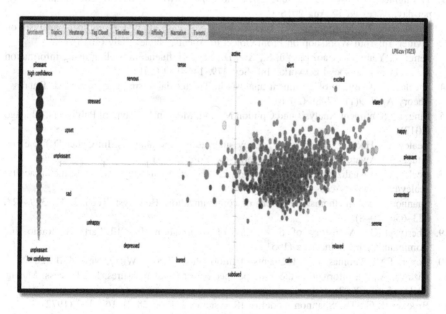

Fig. 7. General sentiments of the select institution

5 Conclusions

Based on the information theoretic review of the study, tweeter feeds can be visualized in a form of charts, graphs and diagrams. Understanding and extracting feelings or emotions from messages or tweeter feeds can also be performed through sentiment analysis. The tf-idf calculated for each query term in tweeter feeds were converted into images using information theoretic approach. With this, the nature of activity and opinions of the users in a select institution were discovered.

In the analysis performed over 1,557 tweets, there was 0.93 mean frequency of tweets posted each day and about 655 user mentions per tweet, 1087 hashtags with 0.70 average number of hashtags, 83 retweets or 5% retweets and 361 replies or 23% replies in the total of analyzed tweets. It showed that the users has more interaction with each other as the numbers calculated increases. As to the query term "Enrollment" used as the target of sentiment, there were 6 tweet feeds analyzed for estimated sentiment. It was visualized that the tweets lie on the right side of the scatterplot diagram which indicates that the tweets as to query term *"Enrollment"* were generally pleasant.

References

1. T. Engineering: Twitter, 30 June 2011. https://blog.twitter.com/2011/200-million-tweets-per-day. Accessed 17 Aug 2016
2. Fornacciari, P., Mordonini, M., Tomaiuolo, M.: A case study for sentiment analysis on Twitter. In: 16th Workshop on From Object to Agents, Naples, Italy (2015)
3. Peng, S., Yang, A., Cao, L., Yu, S., Xie, D.: Social influence modeling using information theory in mobile social networks. Inf. Sci. **379**, 146–159 (2017)
4. Sheela, L.: A review of sentiment analysis in Twitter data using hadoop. Int. J. Database Theory Appl. **9**(1), 77–86 (2016)
5. Bing, L.: Sentiment Analysis and Opinion Mining. Morgan & Claypool Publishers, Chicago (2012)
6. Healey, C., Hao, L., Hutchinson, S.E.: Visualizations and analysts. In: Cyber Defense and Situational Awareness, p. 329. Springer, New York (2014)
7. Healey, C.: Visualizing Twitter Sentiment, 22 May 2016. https://www.csc.ncsu.edu/faculty/healey/tweet_viz/. Accessed 17 Aug 2016
8. Shannon, C.: A mathematical theory of communication. Bell Syst. Tech. J. **27**, 379–423, 623–656 (1948)
9. Cherry, E.C.: A history of the theory of information. In: IEE-Part III: Radio and Communication Engineering (1951)
10. Cover, T.M., Thomas, J.A.: Elements of Information Theory. Wiley, New York (2012)
11. Aizawa, A.: An information-theoretic perspective of tf–idf measures. Inf. Process. Manag. **39**(1), 45–65 (2003)
12. Brookes, B.C.: The Shannon model of IR systems. J. Doc. **28**(2), 160–162 (1972)
13. Ebay: A Guide to Buying Voice Changers (2013). http://www.ebay.co.uk/gds/A-Guide-to-Buying-Voice-Changers-/10000000177317588/g.html
14. University of Twente, January 2017. https://www.utwente.nl/cw/theorieenoverzicht/. Accessed 21 Feb 2017

15. Huang, A.: Similarity measures for text document clustering. In: Sixth New Zealand Computer Science Research Student Conference (NZCSRSC2008), Christchurch, New Zealand (2008)
16. Robertson, S.: Understanding inverse document frequency: on theoretical arguments for IDF. J. Doc. **60**(5), 503–520 (2004)
17. S. a. J. M. P. ". t. s. o. t.-i. b. c. s. p. A. S. R. 3. (. 7.-1. Tata
18. Bakshy, E., Hofman, J.H., Mason, W.A., Watts, D.J.: Everyone's an influencer: quantifying influence on Twitter. In: Fourth ACM International Conference on Web Search and Data Mining, Hongkong, China (2011)

Author Index

Printed in the United States
By Bookmasters